Rethinking the Education Mess

DOI: 10.1057/9781137386045

Other Palgrave Pivot titles

Daniel J. Hill and Daniel Whistler: **The Right to Wear Religious Symbols**

Donald Kirk: **Okinawa and Jeju: Bases of Discontent**

Sara Hsu: **Lessons in Sustainable Development from China & Taiwan**

Paola Coletti: **Evidence for Public Policy Design: How to Learn from Best Practices**

Thomas Paul Bonfiglio: **Why Is English Literature? Language and Letters for the Twenty-First Century**

David D. Grafton, Joseph F. Duggan, and Jason Craige Harris (eds): **Christian-Muslim Relations in the Anglican and Lutheran Communions**

Anthony B. Pinn: **What Has the Black Church to Do with Public Life?**

Catherine Conybeare: **The Laughter of Sarah: Biblical Exegesis, Feminist Theory, and the Laughter of Delight**

Peter D. Blair: **Congress's Own Think Tank: Learning from the Legacy of the Office of Technology Assessment (1973–1995)**

Daniel Tröhler: **Pestalozzi and the Educationalization of the World**

Geraldine Vaughan: **The 'Local' Irish in the West of Scotland, 1851–1921**

Matthew Feldman: **Ezra Pound's Fascist Propaganda, 1935–45**

Albert N. Link and John T. Scott: **Bending the Arc of Innovation: Public Support of R&D in Small, Entrepreneurial Firms**

Amir Idris: **Identity, Citizenship, and Violence in Two Sudans: Reimagining a Common Future**

Anshu Saxena Arora: **International Business Realisms: Globalizing Locally Responsive and Internationally Connected Business Disciplines**

G. Douglas Atkins: **T.S. Eliot and the Failure to Connect: Satire and Modern Misunderstandings**

Piero Formica: **Stories of Innovation for the Millennial Generation: The Lynceus Long View**

J. David Alvis and Jason R. Jividen: **Statesmanship and Progressive Reform: An Assessment of Herbert Croly's Abraham Lincoln**

David Munro: **A Guide to SME Financing**

Claudio Giachetti: **Competitive Dynamics in the Mobile Phone Industry**

R. Mark Isaac and Douglas A. Norton: **Just the Facts Ma'am: A Case Study of the Reversal of Corruption in the Los Angeles Police Department**

Huw Macartney: **The Debt Crisis and European Democratic Legitimacy**

Chiara Mio: **Towards a Sustainable University: The Ca' Foscari Experience**

Jordi Cat: **Maxwell, Sutton and the Birth of Color Photography: A Binocular Study**

Nevenko Bartulin: **Honorary Aryans: National–Racial Identity and Protected Jews in the Independent State of Croatia**

Coreen Davis: **State Terrorism and Post-transitional Justice in Argentina: An Analysis of Mega Cause I Trial**

Deborah Lupton: **The Social Worlds of the Unborn**

Shelly McKeown: **Identity, Segregation and Peace-Building in Northern Ireland: A Social Psychological Perspective**

DOI: 10.1057/9781137386045

palgrave▸pivot

Rethinking the Education Mess: A Systems Approach to Education Reform

Ian I. Mitroff

Adjunct Professor, Saybrook University, San Francisco; Adjunct Professor, School of Public Health, St. Louis University; Professor Emeritus, USC, Los Angeles; President, Mitroff Crisis Management

Lindan B. Hill

Assistant Vice President of the Marian University Academy for Teaching and Learning; Leadership, Marian University, Indianapolis, Indiana

Can M. Alpaslan

Associate Professor, College of Business & Economics, California State University, Northridge

palgrave macmillan

DOI: 10.1057/9781137386045

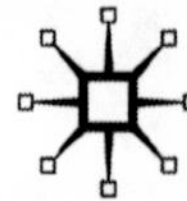

RETHINKING THE EDUCATION MESS

First published in 2013 by
PALGRAVE MACMILLAN®
in the United States—a division of St. Martin's Press LLC,
175 Fifth Avenue, New York, NY 10010.

Where this book is distributed in the UK, Europe and the rest of the world, this is by Palgrave Macmillan, a division of Macmillan Publishers Limited, registered in England, company number 785998, of Houndmills, Basingstoke, Hampshire RG21 6XS.

Palgrave Macmillan is the global academic imprint of the above companies and has companies and representatives throughout the world.

Palgrave® and Macmillan® are registered trademarks in the United States, the United Kingdom, Europe and other countries.

ISBN: 978-1-137-38601-4 EPUB
ISBN: 978-1-137-38604-5 PDF
ISBN: 978-1-137-38482-9 Hardback

Library of Congress Cataloging-in-Publication Data is available from the Library of Congress.

A catalogue record of the book is available from the British Library.

First edition: 2013

www.palgrave.com/pivot

DOI: 10.1057/9781137386045

"We don't see things as they are, we see things as we are."

Anais Nin

"I do not say... that schools can solve the problems of poverty, alienation, and family disintegration. But schools can *respond* to them. And they can do this because there are people in them, because these people [sic] are concerned with more than algebra lessons or modern Japanese history, and because these people can identify not only one's level of competence in algebra but one's level of rage and confusion and depression. I am talking here about children as they really come to us, not children who are invented to show us how computers may enrich their lives..." [italics in original][1]

Neil Postman

Note

1 Postman, Neil, *The End of Education, Redefining the Value of Education*, Vintage Books, New York, 1995, p. 48.

DOI: 10.1057/9781137386045

Contents

DOI: 10.1057/9781137386045

List of Figures and Tables

Figures

Tables

DOI: 10.1057/9781137386045

Preface

This book is about a particular and very important mess: The Education Mess.

To the best of our knowledge, this is the first book to treat education as a complex, messy system. As such, it goes entirely against the grain of the vast, overwhelming majority of books that treat education as if it were nothing but a "machine." In this view, the component parts of education not only exist independently of one another, but as a result, they can be analyzed independently.

As far as we know, Russell Ackoff was the first person to appropriate the term "mess" to stand for a dynamic, constantly changing *system* of problems that are so highly interconnected and bound together such that they can't be separated either in principle, practice, or, most fundamental of all, in actual fact, i.e., their basic existence. Taking any problem out of the mess of which it is a part, not only distorts the basic nature of a problem, but the entire mess as well. The inescapable conclusion is that one deals with messes *as messes* or one has no hope of dealing with them at all.

Education especially calls out for treatment as a "mess" precisely because that's not only what education is, but at the same time, it has not been treated as such. Its very "messiness" has somewhat paradoxically discouraged its being treated as a "mess."

As Ackoff put it:

> [People] are not confronted with problems that are independent of each other, but with dynamic situations that consist of complex systems of changing problems that interact with each other. I call such situations *messes.* [emphasis in original]

DOI: 10.1057/9781137386045

> Problems are abstractions extracted from messes by analysis.
>
> Therefore, when a mess, which is a system of problems, is taken apart, [i.e., analyzed] it loses its essential properties and so does each of its parts. The behavior of a mess depends more on how the treatment of its parts interact than how they act independently of each other. *A partial solution to a whole system of problems is better than whole solutions of each of its parts taken separately.*[1] [emphasis ours]

No matter what one believes its underlying causes are, and therefore how best to deal with it, many accept the proposition that education is a mess. Indeed, statements to the effect that education is a mess are not uncommon. However, once having said this, agreement as to what to do quickly vanishes. More importantly, as we noted earlier, there are virtually no in-depth treatments of education as a mess.

The extreme divergence and abject bitterness between different philosophical positions, values, and worldviews about what to do to "'solve' The Education Mess" (hereafter referred to as TEM) quickly take over and dominate the debate. In fact, it quickly becomes apparent that different parties don't see the "same mess" to begin with, let alone whether it's "solvable or not." Even more basic, virtually no one except Ackoff goes beyond using the word "mess" in any but the most pejorative and deprecating manner of speech. Presumably, once something has been labeled a mess, little if anything can be done about it. Apparently, the only thing one can do is to throw up one's hands and slink away.

This book not only shows how to represent, and thereby, better understand, TEM *as a mess*, but it also provides key heuristics for coping with it. We not only show how and why education is a mess, but as a result, why it does not have nice, neat, and exact solutions like the more often than not simple and misleading exercises that are typically found at the end of most texts. In fact, we show that the processes of "representing" and "coping with" messes are not independent activities.

One of the most powerful ways of understanding messes is through our coping with and attempting to manage them. Often, it is the only way of understanding them. Only exercises offer a complete definition of the problem prior to our working on them. In contrast, a definition of a mess only emerges, if then, through the process of working on and attempting to manage it. As opposed to exercises, the order of definition is completely reversed with regard to messes. But since people are generally miseducated through a long process of being fed exercises, they are generally thrown for a complete loss if they are not given a precise and

DOI: 10.1057/9781137386045

unvarying definition of a problem before they start work on it. Indeed, for most people, this process of miseducation begins early and ends late, if at all.

Exercises also go hand in hand with reductionism and empiricism. This does not mean that we are opposed to collecting data and subjecting our ideas to strict empirical tests. Indeed, we insist on it. Instead, the concept of a mess leads to a different form of empiricism.

Like viruses, TEM can only be "coped with" and "managed," never "'completely and finally solved or eliminated' in its entirety." This does not mean that hopelessness and despair are inevitable.

If messes are not like artificial, simplistic exercises, the fact that there are heuristics for coping with them shows that there are ways of managing them. But this is possible if and only if we recognize and finally accept that all of the complex problems of modern societies are not only parts of messes, but as a result, can only be properly addressed as messes, not as self-standing, independent problems.

One of the most powerful ways of showing this is by means of a very special form of systems thinking that is based on the pioneering work of Carl Jung. Mitroff and a life-long friend and colleague Ralph Kilmann developed it in the early 1970s when both were at the University of Pittsburgh.

Finally, we have tried to make the ideas as accessible and as clear to the widest possible audiences. For this reason, we hope that it can be fruitfully read and used by as many people as possible that affect and are affected by the state of education. This includes, but is not limited to, professors of education, classroom teachers, policy makers, parents, principals, superintendents, and the officials of teacher's unions. Since we believe that one is never too young or too old to learn about systems, our hope is that students will also be motivated to learn about TEM as well. In sum, it is for everyone that is impacted by education. In today's world, that includes virtually everyone.

Note

1 Ackoff, Russell L., *Re-Creating the Corporation*, Oxford University Press, New York, 1999, pp. 178–179.

DOI: 10.1057/9781137386045

Outline of the Book

Chapter 1 gives a brief introduction and summary of the main issues. In particular, it discusses the extreme disagreement and bitter acrimony between the proponents of new, charter schools and traditional public schools.

Chapters 2 and 3 are foundational. They set out the basic concepts and ideas about systems and messes that we use throughout. They are vital if one is to be able to even think about thorny and difficult problems such as TEM.

Chapter 4 presents an analysis of charters from a very different form of system's analysis. In particular, we show how a different conception and use of the Jungian personality framework can not only be used for system's analysis, but gives a deeper and richer insight into those few, special charters that are highly successful in lowering the achievement gap between (1) mostly middle and affluent, upper class White students and (2) White, Black and Hispanic, poorer urban students.

Chapter 5 primarily examines the role of teachers and unions. It also takes a look at possible designs for the charters of the future.

Chapter 6 looks at planning efforts that are underway in Indianapolis and the state of Indiana to lower the achievement gap. It argues that as admirable as these efforts are, they are deficient in terms of the analysis presented in Chapter 4. Unfortunately, this same criticism applies to the Community Schooling Movement however admirable it is on many other grounds.

DOI: 10.1057/9781137386045

Chapter 7 summarizes a set of key heuristics for coping with messes. It argues that the model laid out in Chapter 4 is key in using the heuristics to cope with The Education Mess.

As this book was in its final stages of completion, the tragic shootings at Sandy Hook Elementary School in Newtown Connecticut in 2012 occurred. The tragedy pushed us to take a deeper and more serious look at school shootings and violence in general as one of the most important parts of TEM. Chapter 8 thus applies the heuristics of Chapter 7 in grappling with school violence and security more broadly than they have typically been dealt. To do this, we not only consider school violence and security from the perspectives of psychology and sociology, but from the deeper perspective of psychodynamics as well. We thus go at it as it were from the "outside-in." That is, we go from more easily observable external factors and forces to deeply internal factors/forces that are less easily and unfortunately seldom observed. As we show, both sets of factors/forces are needed to make sense, if one ever can, of the senseless. The result is a broader treatment of mental health issues that affect school safety and security.

Messes always have a high potential for crises. In Chapter 9, we discuss various aspects of Crisis Management in the context of schools. For Crisis Management is not only an indispensable skill. It is a vital part of the necessary mindset to treat and cope with messes. In short, it is an integral component of our ability to manage TEM, and by the same token, any other mess.

Finally, in the Epilogue, we summarize very briefly some of the many themes of the book.

DOI: 10.1057/9781137386045

About the Authors

Ian I. Mitroff is widely regarded as one of the "fathers" of modern Crisis Management.

He has spent his entire career bringing interdisciplinary approaches in order to find successful solutions to complex issues.

In 2006, he became Emeritus Professor at the University of Southern California (USC) where he taught for 26 years. While he was at USC, he was Harold Quinton Distinguished Professor of Business Policy in the Marshall School of Business; he also held a joint appointment in the Department of Journalism in the Annenberg School for Communication at USC where he taught Crisis Management, and where he was also Associate Director of the USC Center for Strategic Public Relations.

Currently, he is Adjunct Professor at Saybrook University, San Francisco, California. He is also an Adjunct Professor in the School of Public Health at St. Louis University, St. Louis, Mo. He is Senior Research Associate at the Center for Catastrophic Risk Management at the University of California at Berkeley.

Mitroff has a Ph.D. in Engineering Science (Industrial Engineering) and a minor in the Philosophy of Social Systems Science from UC Berkeley. He is Fellow of the American Psychological Association, the American Association for the Advancement of Science, and the American Academy of Management. He has an honorary Ph.D. from the Faculty of Social Sciences at the University of Stockholm. He is the recipient of a gold medal from the United Kingdom Systems Society for his life-long contributions to understanding complex problems. The award was presented in September 2006 at St. Anne's College, Oxford University.

For 35 years he has been sought out as an analyst and consultant with regard to human-induced crises, including such major incidents such as the Tylenol poisonings, Bhopal, Three Mile Island, 9/11, the scandal in the Catholic Church, Enron, the war in Iraq, and the Tsunami in Southeast Asia.

Mitroff is also the author of several well-received books, including *The Unreality Industry*, co-written with Warren Bennis (1989), *A Spiritual Audit of Corporate America* (1999), *Crisis Leadership* (2003), and *Why Some Companies Emerge Stronger and Better From a Crisis* (2005).

His most recent books are *Dirty Rotten Strategies: How We Trick Ourselves and Others into Solving the Wrong Problems Precisely* (with Abe Silvers, 2009), and *Swans, Swine, and Swindlers: Coping with the Growing Threat of Mega-Crises and Mega-Messes* (with Can Alpaslan, 2011).

Mitroff is also president of his own private consulting firm, Mitroff Crisis Management, which offers an integrated approach to Crisis Management. His past clients have included *Fortune* 500 Companies, governmental agencies, and not-for-profit organizations.

Lindan B. Hill is currently Dean of the School of Education and Director of the Marian Academy for Teaching and Learning Leadership at Marian University, Indianapolis, Indiana. He has been Dean since June, 2006. Lindan received a bachelor's degree in English from Indiana University in 1969. Upon receiving his BA, he taught in the neediest section of the inner city in Miami, Florida. Returning to Indiana, Lindan received his Master and Doctor of Philosophy degrees from Purdue University and continued his public school career as teacher, alternative school director, secondary school principal and, for 25 years, superintendent of schools in two districts in Indiana.

Upon retirement from public schools, he served as Director of Teacher Education at Manchester University before becoming Dean of the School of Education at Marian University. Lindan has served on a number of professional committees and advisory groups, including the Charter School Advisory Board for Mayor of Indianapolis, United States Department of Education National Blue Ribbon School Selection Committee and President of the Indiana Association of Public School Superintendents.

DOI: 10.1057/9781137386045

In 1996, Lindan was awarded the Distinguished Hoosier designation from Governor Evan Bayh. In 2001, he was awarded Indiana's highest civilian award, Sagamore of the Wabash, from Governor Frank O'Bannon.

Can M. Alpaslan is Associate Professor of Management in the College of Business and Economics at California State University, Northridge. He has a Ph.D. from the Marshall School of Business at the University of Southern California. He also holds a MBA degree from Bilkent University, and a B.Sc. in Mechanical Engineering from the Middle East Technical University.

He is a member of the Academy of Management and the Society for Business Ethics. He is an editorial board member of the *Academy of Management Learning and Education*. His articles have been published in journals such as *Harvard Business Review*, *Journal of Contingencies and Crisis Management*, *Management Communication Quarterly*, *Journal of Management Education*, and *Journal of Management Inquiry*. He is the author of *Swans, Swine, and Swindlers: Coping with the Growing Threat of Mega-Crises and Mega-Messes* (with Ian Mitroff, 2011).

Alpaslan has worked closely with Mitroff in conducting research on Crisis Management.

DOI: 10.1057/9781137386045

1
Introduction—TEM, The Education Mess

Abstract: *This chapter argues that education is a mess. A mess is a system of problems that are so highly interconnected such that no problem exists or can be studied independently of all the other problems that constitute the mess and the entire mess itself. The chapter also lays out the debate with regard to what to do about TEM, The Education Mess. On the one side are those who favor charter schools. They believe that the public schools have failed and therefore need to be radically redesigned, if not jettisoned altogether. On the other side are those who support public schools and seek to improve them, not abandon them. The chapter argues that both sides need one another more than they realize. Both are needed if we are to have any hope of coping with TEM.*

Mitroff, Ian I., Hill, Lindan B., and Alpaslan, Can M. *Rethinking the Education Mess: A Systems Approach to Education Reform*. New York: Palgrave Macmillan, 2013.
DOI: 10.1057/9781137386045.

"'... People need to think a little more about the problems kids in this school [Manual High School in Indianapolis, Indiana] have and the issues they have to deal with day in and day out [alcoholism, bureaucratic and uncaring administrators and teachers, chronic poverty, constant threat of crime and violence, drugs, divorce, guns, homelessness, low parental and teacher expectations, school and parental apathy, parental abandonment, teenage pregnancy, etc.]. There are a lot of social issues and a lot of home drama. There are a lot of things, a lot of factors that go into a school being unsuccessful. People want to say it's the kids. Or the parents. Or the teachers. Or the system. [sic] It's not that easy. There's no one factor that can turn everything around. Americans want quick fixes and easy solutions. Sorry, there isn't one when it comes to education. All you can do ... is put every ounce of energy you have into helping every student you can.'"[1]

Rich Haton [Teacher in Manual High School]

Introduction

A central theme of this book is that education is a mess. While many, if not virtually all, who study education agree in some form or another with the broad proposition that education is a mess, having said this, agreement quickly vanishes. The extreme divergence and abject bitterness between different philosophical positions, values, and worldviews about what to do to "'solve' TEM" quickly take over and dominate the debate. Indeed, different parties don't see the "same mess" to begin with, let alone whether it's "solvable or not."

The conflict is so bitter, deep, and intense that it virtually prevents—paralyzes—everyone from seeing that the solution does not lie in any of the one-sided perspectives and extreme worldviews. Rather, if there is a "solution," it consists in forming new perspectives and worldviews that integrate and go far beyond the old ones. More than they realize, all positions are highly dependent upon one another.

In fact, we show that none of the previously stated positions can even define the problem adequately, let alone solve it, acting solely by themselves. They need to incorporate seriously the very things to which they are so strongly opposed in order to form richer definitions of "the

DOI: 10.1057/9781137386045

problem." In short, they sorely need richer definitions before they can ever hope to find "solutions."

In brief, the extreme divergence between positions and worldviews is itself one of the biggest, untreated contributors to TEM. While discussion of this particular aspect of TEM has of course not been absent altogether, it has not been given center stage and thus the full treatment it demands. For all practical and theoretical purposes, the proponents of different positions live in parallel universes. They are not merely "ships passing in the night," but are more like "far flung galaxies billions of light-years apart."

Although this book is not solely about charter schools, the charter school movement is one of the best issues to illustrate the extreme divergence between views regarding the state of K-12 education in America and how to improve it. And, although The Charter School Mess (TCSM) is not completely equivalent to TEM, they are close enough for our purposes such that treating TCSM allows us to treat TEM. In fact, TCSM is so deeply intertwined with TEM that they cannot be treated separately.

Charter schools

Briefly, a charter school is a publicly funded school that is normally governed by a group or organization under a legislative contract with a particular state. The charter exempts the school from certain local and/or state rules and regulations. In return for their independence, a charter school must meet strict accountability standards that are laid down in its initial contract. Normally, a school's charter is reviewed every three to five years to see whether it is following specified guidelines on curriculum and management, and whether certain standards such as the marked improvement in student scores on standardized tests, typically reading and math, are being achieved. If it is not, then there are strong grounds for a charter's closure.

At the present time, there are approximately 55,000,000 students in public schools. In comparison, it is estimated that there are only 1,500,000 students in charter schools, or roughly 2.7% of the public school population. While small in numbers, the charter school movement is nonetheless huge in its implications.

To say that the proponents and the critics of charter schools are divided, if not as a general rule extremely hostile to one another, is

DOI: 10.1057/9781137386045

putting it mildly. More often than not, it seems as if they truly despise one another.

The reformers: pro-charters

On the one side of the debate about charters are The Reformers. For them, the current public school system is irredeemably broken. It cannot be fixed, period! The only hope lies in rebuilding the system around entirely new kinds of schools, that is, charter schools and the like.[2]

To support their argument, the Reformers repeatedly trot out the fact that the current system of public schools has failed miserably to make a dent in the persistent achievement gap in math and reading scores between (1) poor, urban, mostly Black and Hispanic children, and (2), mostly white, middle and upper middle-class children. (Upper middle-class and upper class children already opt out of public schools to a large and growing extent by going to costly private schools.) Freeing charter schools from the bloated bureaucracies, archaic teacher unions, and underlying attitudes of the current system ("poor kids don't have what it takes to succeed") that do more to protect adults than they do to help children, and exposing them to dedicated, outstanding teachers is the only thing that can save all children.

Charter schools can do this because unlike traditional public schools, they are free to attract and hire the best teachers, and fire them if they don't perform. They are not burdened with the constraints of the current public school system that forces them to accept whatever teachers a district imposes on them. They are also not bound by the seniority rules of public schools that insist that preferences in hiring be given to older teachers even if younger, and other older, teachers are better, more qualified, etc.

Charters are also generally in favor of testing children and basing teacher evaluations, promotion, and retention on student performance. In short, teachers are judged primarily on how well they do in getting children to pass standardized tests. They are also judged on a host of other factors such as encouraging creativity and in motivating children to want to succeed.

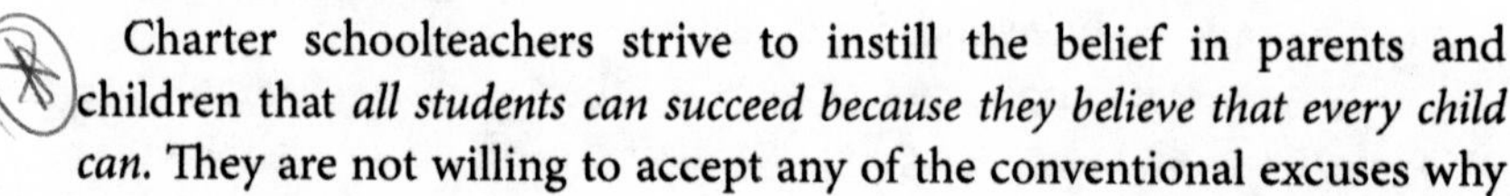

Charter schoolteachers strive to instill the belief in parents and children that *all students can succeed because they believe that every child can.* They are not willing to accept any of the conventional excuses why

DOI: 10.1057/9781137386045

children cannot learn, that is, the general problems of society, poverty, poor health and living conditions, etc.

Public School Advocates: anti charters

Public School Advocates are on the other side of the debate.[3] They are often mistakenly characterized as anti-reformers, which they are not. Public School Advocates believe in public schools because they believe that it is one of the basic duties of a democratic society to educate all of its children in a shared public setting where they get to know, and hopefully respect, one another through prolonged and intensive interaction. Only in this way can they become citizens who can empathize with and relate strongly to people that are different from them.

We hope that it will become clear that although we are very strongly in favor of a highly select, special set of charters, we agree with the view that children need to be exposed to and interact with others that are different from them. We are also very strongly in favor of *not* abandoning public schools altogether. But make no mistake about it. We believe that it is not business as usual. Public schools have to change, and not in minor, but in major ways. But this is equivalent to saying that the general public also has to change in major ways in its understanding and support of public schools, if not all schools. In this sense, we do not regard charters as the "solution" or "final model." Rather, we regard charters as an "educational innovation lab for the schools of the future."

Public schools are under a tremendous burden and disadvantage because unlike charter schools, they cannot "cherry pick" motivated parents and students who are willing to do the extra, hard work that it takes to learn, and thereby to close the achievement gap. Public schools have to accept everyone in their surrounding district, rich and poor, advantaged and disadvantaged, abled and disabled, etc.

This particular criticism of charters is less true today than it was. It is certainly not true of Indianapolis, the city we know best, where the population of students in charters is very mixed and heterogeneous that it essentially matches the population of the surrounding public schools.

Public School Advocates are generally opposed to testing because teaching to tests quickly becomes the norm and thus gets in the way of education. In their view, standardized tests are more often than not unreliable and invalid. They do not give a true assessment and picture of

DOI: 10.1057/9781137386045

what education is about, that is, critical thinking, being a good citizen, life-long learning, etc. Tests promote cheating and gaming. Linking teacher performance solely or mainly to tests is a grievous distortion of what good and great teachers strive to accomplish. Further, linking merit pay to how well students do on tests not only promotes divisiveness between colleagues, and hence destroys collegiality, but it is also intrinsically fallacious because who is a "good" or "great" teacher varies from class to class and year to year.

Public School Advocates also believe that what happens *away* from school is as important, if not even more so, than what happens *at* school. *External or Outside* factors such as crime, hunger, low-wage jobs of parents, poor health, poverty, poor living conditions, etc. are as important, and even more important, than *Internal or Inside* factors at school, that is, class size, school conditions, leadership, quality of teachers, etc. Thus, the way to improve education is to improve the general conditions of society, strengthen school curricula, raise the prestige of teachers, etc.

Public School Advocates also argue that as successful as some charters may be, there is no practical way to scale up the results to make them applicable to all schools. In short, charters operate under a very special set of non-replicating situations. What happens in the small cannot always work in the large.

The debate charges on

The proponents of charters and special schools argue that they are not willing to wait for society to solve all its ills and problems before the achievement gap is lowered. They've proved that it can be done.

In turn, Public School Advocates respond with their own counter argument: if you look at the hard data that Reformers advocate so strongly as the final arbiter or measure of progress, then there is little overall difference between the general performance of charters and public schools in average math and reading test scores of students.[4] There is of course more than a modicum of irony in Public School Advocates using test data to judge charters since Public School Advocates don't generally believe in the use of standardized tests—or at least not solely—to evaluate something so complex as teaching and learning. But then, in their view, charters have so-to-speak laid down the "initial data challenge," not Public School Advocates.

DOI: 10.1057/9781137386045

Both sides are profoundly right and wrong. Special charters such as The Harlem Children's Zone have shown that with a very distinct and highly integrated set of surrounding support services and systems (health, housing, jobs, parent training, etc.), plus extremely dedicated and motivated parents and children, poor children can succeed.[5] In other words, *one very special type of social experiment* shows that it *is* possible to lower the achievement gap significantly. In this sense, the Reformers are right. But so are the Public School Advocates. The Harlem Children's Zone is a success only because it contains a highly orchestrated set of special social programs that bolster the health, housing, and general welfare of poor parents and families. In this way, children come to school more able and willing to learn. The Harlem Children's Zone has thus accepted and incorporated one of the most important principles of some of the more ardent Public School Advocates. Namely, external factors—what goes on outside of schools—are as important as internal factors—what goes on inside.

Again, although we are generally in favor of charters, we believe that no charter should even be considered, let alone allowed to operate, unless it contains a serious plan for community involvement and improvement. In systems terms, what goes on *outside* and *inside* of schools are *inseparable. At the very least, a charter must have a long-range plan for involving and improving the community over time.*

More highly dependent than they realize

The upshot is that the Reformers and Public School Advocates are more highly dependent upon one another than they have dared imagine in their wildest dreams, or better yet, nightmares. The remainder of this book explores the implications of this and the new forms of alternative and public schools that we have just begun to imagine, and that are sorely needed if we are truly to close the achievement gap by improving all schools.

Concluding remarks: complex, messy systems

The broadest issue with which we are concerned is complex, messy systems and what they say about the design and operation of schools in

DOI: 10.1057/9781137386045

the 21st century and beyond. To do this, we have to examine the nature of problems, systems, and messes. We especially have to examine the psychological and philosophical underpinnings of very different views of education. We have to get beneath the surface of the arguments pro and con that currently bog down education. If the debate over charters has proven anything, it won't be settled by the usual ways in which we attempt to settle serious arguments, i.e., by more studies and better data *alone*.

The key word is "alone."

It is not because more good studies and data are not needed. Of course they are. We always need all the good studies and data we can get. Indeed, as we show, we need entirely new kinds of studies and data.

But studies and data alone are generally insufficient to move people out of deeply held positions. What's needed are better means of exposing and overcoming deeply held, and often unconscious, assumptions about the world.

A joke that preeminent systems philosopher Russell Ackoff often told is relevant here: A man goes to a psychiatrist, who asks, "How can I help you? What's your problem?" The man replies, "I'm dead, but I can't get my friends and family to accept it." The psychiatrist nods and says, "Uh-huh."

After six months of working with the man, the psychiatrist asks, "Look if I can prove to you that you're not dead, will you give up the belief that you are?" "Of course," says the man. The psychiatrist then asks, "You don't believe that dead men bleed do you?" "Of course not! That's impossible!" With this, the psychiatrist takes a small pin out of his desk and pokes the man such that a small drop of blood appears on his arm. The man looks at the small drop and exclaims, "Well, I'll be darned! Dead men do bleed!"

The point is that without knowing the origin and purpose of the man's belief system, all the data and arguments *alone* are generally powerless to change anyone's beliefs and behavior. Go try and change the beliefs of fervent pro-life proponents versus pro-choice by data and arguments alone. The same applies to the beliefs of fervent gun proponents versus those who want tougher guns laws and restrictions. We take the latter issue up in more detail in the last chapter with regard to greater school safety.

Finally, we need to make clear that this book is not a detailed plan for coping with TEM. It does not specify in detail the wide range of actions

DOI: 10.1057/9781137386045

and strategies that are necessary to deal with the many stakeholders that impact and are impacted by education. Instead, it is a broad roadmap of how to conceptualize the many factors and forces that impinge on TEM, if not on all messes.

If we have added to our basic knowledge of how to think about messes, then we will have achieved our primary goal. Above all, we hope one thing will become abundantly clear. From start to finish—there really is no "finish"—TEM is a mess. It must be conceptualized and coped with as such if we are to have any hope of managing it.

None of this should be interpreted as saying that we do not offer strong prescriptions about TEM. In fact, we offer a number. Nonetheless, our primary intent is to encourage others to go beyond the ideas that we have learned about messes.

Notes

1 Tully, Matthew, *Searching for Hope: Life at a Failing School in the Heart of America*, Indiana University Press, Bloomington, 2012, p. 18.

2 For some of the many proponents of this world-view, see Merseth, Katherine, *Inside Urban Charter Schools*, Harvard Education Press, Cambridge, Mass., 2009; Thernstrom, Abigal and Stephan, *No Excuses: Closing the Racial Gap in Learning*, Simon & Schuster, New York, 2003.

3 For some of the many proponents of this world-view, see Ravitch, Diane, *The Life and Death of the Great American School System: How Testing and Choice Are Undermining Education*, Basic Books, New York, 2010; Rothstein, Richard, *Grading Education: Getting Accountability Right*, EPI, Washington, DC, 2008.

4 Stuit, David, and Stringfield, Sam, "Special Issue: Responding to the Chronic Crisis in Education: The Evolution of the School Turnaround Mandate," *Journal of Education of Students Placed at Risk*, Vol. 17, Nos 1–2, January–June, 2012.

5 Dobbie, Will and Fryer, Roland, "Are High Quality Schools Enough to Close the Achievement Gap? Evidence from a Social Experiment in Harlem," *National Bureau of Economic Research*, Cambridge, Mass., 2009; Tough, Paul, *Whatever It Takes, Geoffrey Canada's Quest to Change Harlem and America*, Mariner, New York, 2009.

DOI: 10.1057/9781137386045

2

What Is a System and What Is a Mess?

Abstract: *This chapter and the next introduce the basic concepts and ideas that are used throughout to give a very different kind of analysis of TEM and eventually to offer strong prescriptions regarding how to "cope" with it. Since the concept of a mess is dependent on the concept of a system, we first define the concept of a system, especially social systems. The result of the discussion is that as important as it obviously is of having outstanding dedicated teachers, by itself no single component of a mess can improve the total performance of a system or mess.*

Mitroff, Ian I., Hill, Lindan B., and Alpaslan, Can M. *Rethinking the Education Mess: A Systems Approach to Education Reform*. New York: Palgrave Macmillan, 2013. DOI: 10.1057/9781137386045.

 DOI: 10.1057/9781137386045

"If poverty is a disease that infects an entire community in the form of unemployment and violence, failing schools and broken homes, then we can't just treat those symptoms in isolation. We have to heal that entire community. And we have to focus on what actually works."[1]

US President Barack Obama

"The problems that education faces today are systemic problems. They are deeply embedded within the structures under which schools operate. They are deeply embedded in family and social problems. They have to do with the other systems that operate in the broader environment in which schools themselves exist, what we call the macroenvironment [sic]. Schools cannot change through programs or partnerships alone. They can change only if they take charge of themselves and begin to rethink their systems, their internal operating procedures, and the ways in which they relate to those constituencies that they once considered 'outside.'"[2]

Sandra Waddock

Introduction

In this chapter, we primarily confine our discussion to The Charter School Mess (TCSM). The reason is that TCSM is "messy" enough such that it incorporates many, if not all, of the same thorny issues and problems that plague TEM. Indeed, one type of an educational mess relates directly to others. In fact, as we shall see *ALL* messes not only relate to, but are parts of one another. (In Chapter 8, we expand our treatment of TEM by treating one of its most unfortunate components, school shootings and violence, which are messes themselves.)

We begin with the most fundamental concepts: systems and messes. Since a mess is defined in terms of a system, we thereby start with the concept of a system. But there is another important reason why we start with a system.

Like most fields, education is generally confused with regard to the meaning of a system. A system is more than a multitude of stakeholders or boxes with an entangled web of arrows going back and forth between them.[3] It is also more than the multiple dependencies and interactions between manifold entities such as local schools, school districts, State

DOI: 10.1057/9781137386045

Departments of Education, the Federal Department of Education, unions, etc. To be sure, a system is certainly this, but it is much, much more as well.

In *Not By Schools Alone, Sharing Responsibility for America's Education Reform*[4], Sandra Waddock is essentially the only other writer that develops a systems' view of education that is extremely close to the one developed in this book. The main difference is that we have a different view of systems and thus reach different but compatible conclusions. It is not incidental that Dr. Waddock is a Professor of Business Strategy, i.e., from "outside" of education.

Because the discussion of systems and messes can very easily and quickly becomes highly abstract and complex, we want to ground it as much as possible by using an example from the field of education. We make constant reference to this example throughout this and the next chapter. We also expand upon it as we proceed to show that not only does it have all the features of a system, but of a mess as well. For the sake of clarity, we use other examples from outside of the field education as well.

The Turnaround Movement

The term "turnaround" gained traction in education when after 9/11 Harold Levy, Chancellor of the New York City school system, asked Pamela Cantor, the President of Turnaround for Children, to work with children who were in crisis.[5] Duke captures the essence of the movement as follows:

> By December 2009, the Department of Education ... worked out the specifics of ... turnaround guidelines. States and districts were required to target Title I schools that ranked in the bottom 5% in student achievement. The definition of *lowest-achieving schools* for the first time was expanded to include high schools with graduation rates consistently below 60%. School districts were presented with a short list of four options for achieving significant improvements in lowest-performing schools.
>
> Option 1 called for closing the school and sending students to higher achieving schools. Option 2, the so-called *restart model*, involved turning operation of the school over to a charter or education-management organization. The *transformation model*, option 3, required replacing the principal and improving the school through comprehensive curriculum reform,

DOI: 10.1057/9781137386045

> professional development, extended learning time, and other reforms. The final option [4], dubbed the *turnaround model*, entailed replacing the principal, screening existing school staff and rehiring no more than half the teachers, choosing a new governance structure, and adopting reforms similar to the transformation model. Thus, for the first time, at least for policy purposes, the term *turnaround* came to be associated with a relatively specific set of changes, not just the notion of a quick and dramatic improvement in student achievement [italics in original].[6]

As we shall see very shortly, The Turnaround Movement exhibits all of the prime features of systems and messes.

What is a system?

In a series of seminal books spanning a lifetime, no one has done a more commanding and effective job than Russell L. Ackoff and his colleagues in identifying and laying out the precise definition and nature of systems, especially *social* systems.[7]

First of all, *ideally a system is an intentionally designed, systematically organized, whole entity (e.g., an automobile, computer, smart building, etc.) that has one or more essential functions so that an individual and/or group of people are thereby able to realize a set of important purposes.* (It is important to note that this definition does *not* exclude systems that are the result of unintended actions and consequences, for the key word is "ideally." Part of the reason why systems are so complex is that they include both intentional and unintentional actions and consequences.) *Furthermore, the functions, not the parts, are critical in defining the system.* (Notice that the "group of people" and their "associated purposes" may in turn constitute additional systems.)

For example, in terms of systems thinking, an automobile is defined primarily in terms of its function(s), not its parts, although the parts are certainly critical, for without them, the functions cannot be accomplished.

A car's functions are to allow people to accomplish specific purposes, i.e., move to a series of desired locations by means of desired routes in desired times. Cars also have additional functions such as to enable people to engage in entertainment and relaxation, although more often than not, increasing traffic congestion defeats these goals.

A critical distinction is that a system's parts have functions *while only humans as purposive individuals have* purposes. Thus, a car has major

DOI: 10.1057/9781137386045

functions (e.g. the abilities to change direction and speed when directed by a purposeful being, etc.) that allow humans to satisfy purposes in the form of desired outcomes (e.g., get to a specific location in a certain time, serve as a status symbol, etc.).

The crucial difference is that only humans purposefully create means and ends and then choose specific means to accomplish particular, intended ends. In this way, humans display and engage in purposeful actions and beliefs. In contrast, systems and machines carry out predetermined, designed *functions* (means) made possible by parts that should be carefully designed in order to accomplish intended *purposes* (ends). In brief, *only humans* (and of course certain other animals) *are purposeful beings and thus exhibit purposive behavior* even if they are not completely autonomous and self-contained. In other words, only humans and other animals are capable of intentional behavior.

(Of course, with the advent of machines that emulate human thinking, the lines between humans and machines have been muddied considerably. And, they will only get muddier with at first the placing of computer chips in people to monitor health, and then later to augment and improve on human abilities. Whether we like it or not, we are already well on the way to becoming Cyborgs!)

Individual humans are not completely autonomous and self-contained because they exist only by virtue of their being members of even larger systems, e.g., families, organizations, and societies. Human infants do not have the innate ability to survive and develop completely on their own. In short, the lines between individuals and the society of which they are members are very thin. (Further, in a world that is increasingly global, the lines between societies have become vanishing small as well.) In fact, neither individuals nor societies exist without the other. In this sense, there are no absolute differences between the sciences of psychology and sociology. From a systems perspective, they never should have developed independently of one another. The same is true of all the social sciences and related fields, e.g., anthropology, economics, history, etc.

To take another example, the heart and lungs have essential functions (pumping and circulating blood and air), but they don't have independent purposes, let alone an existence of their own apart from the entire human body. Similarly, a car engine obviously has an important function, but it doesn't have a purpose of its own independently of the combined human-machine system that directs it to desired ends.

DOI: 10.1057/9781137386045

By themselves, wheels do not exhibit *purposeful* motion. Indeed, by themselves, wheels just roll around, fall down, and finally come to a complete rest. Wheels only carry out their intended function by being part of a car as a whole system that not only includes, but is directed by, a purposeful being.

Given the ever present and persistent human propensity to imbue virtually all things, natural and otherwise, with purposes, it is extremely difficult to keep the distinction between functions and purposes clear and strictly apart. This is especially the case when the "parts" of a system are themselves purposeful individuals, organizations, or institutions. Nonetheless, it is important to hold on to the distinction for it serves a very important "purpose."

In short, we need to keep firmly in mind that functions are not purposes. Functions exist to help us carry out and realize intended purposes.

The Turnaround Movement as a system

The Turnaround Movement exhibits many of the prime features of a system. First, it has been "intentionally designed" to accomplish a specific set of purposes, the most important of which is to turnaround low-performing schools. Indeed, it is the direct intention of a great many purposive individuals and stakeholders: parents, the Federal Government, principals, students, teachers, etc. Whether it is a "well-designed system" is of course another matter, if not the crux of it. Second, the very definition of "lowest-performing schools" is an essential part of the system. In fact, the definition cannot be decoupled from the rest of the system. We will encounter other instances of how it is a system as we proceed.

While the definition of a what a school is like that "needs turning around" does not automatically determine the precise "mechanisms"—options—that have been designed in order to help bring low levels of achievement up to "par," it does set in motion the search for mechanisms that will realize the purpose of turning around low-performing schools. In this regard, the various options play a role similar to that of a car wheels, engine, etc. They embody functions that will hopefully realize the purposes of the system's "designers." In this sense, while they are not purposes per se, they reflect them nonetheless. They are certainly not independent of purposive human designers.

DOI: 10.1057/9781137386045

What is a system continued?

In addition, *a system also consists of at least two or more essential parts that satisfy three conditions.* If something has only one part, then it is not a system.

The first condition is that a system cannot accomplish its defining function(s) without its essential parts. An engine is an essential part for locomotion but a cigarette lighter is not. Of course, a cigarette lighter can be an essential part if smoking is considered to be an important purpose. Similarly, the brain, heart, and lungs are essential parts of humans, but as Ackoff notes, the appendix is not. This is in fact why it is termed an "appendix."

The essential parts of the Turnaround Movement are more than the various options themselves. For instance, in addition to options, the essential parts consist of all of the purposive humans, organizations, and institutions such as Colleges and Professors of Education, researchers, policy makers, parents, students, etc., who have a prime stake in developing and administering tests that define and measure achievement.

The second condition is that by itself an essential part cannot affect a system independently of at least one other essential part. The essential parts are not only interconnected, but they also interact. Thus, the heart affects the lungs and vice versa. Stronger still, they don't even exist without the other. Without interactions and interdependencies, there is no system. This is not only critical to the definition of a system, but even more so for messes.

In The Turnaround Movement, critical parts of the system are additional factors such as the geographical and cultural "proximity" of "better-preforming schools." A critical interaction and thus property of the system as a system is whether parents will opt to disrupt their children's education and immediate classroom friendships by pulling them out of poor schools and sending them to better schools even if better schools are nearby. In this way, the definitions and meanings of what are "better" and "nearby" are crucial parts of the system.

The third condition is that no group of a system's essential parts—that is, no subsystem—has an independent effect on the whole system. Once again, the nervous and metabolic subsystems of humans do not have independent effects on the whole human body as a system. This feature certainly holds true as well for the Turnaround Movement.

These definitions and conditions have important consequences for the performance of systems and thus illuminate additional properties.

DOI: 10.1057/9781137386045

Improvement in the parts taken separately does not necessarily improve a system overall as a whole. Indeed, it often leads to its failure and complete destruction. Merely improving an engine without the careful coordination of and simultaneous improvements in the suspension and transmission does not improve the overall performance of a car. If anything, it can cause a car to spin dangerously out of control.

Clarifying and improving the "exact definition and measurement of low-performing schools" without the ability to actually implement the changes that are necessary to improve such schools does not improve the overall performance of the entire system. This is in fact a general criticism of research and policy making in education as a whole. To be fair, this same criticism applies to virtually all fields of knowledge and professions. It certainly applies to the special collection of articles referred to earlier that identify important factors that pertain to the Turnaround Movement. While nearly all of the individual articles are excellent, there is no single article that shows how all of the separate factors interact systemically.[8] No article gives a systems overview that even indicates how all of the factors that the separate articles identify are highly interdependent.

Healthcare is another highly important example. Attempts to improve the overall costs of medical care by lowering the costs of the individual parts of the system have failed. In fact, they have done just the opposite, i.e., raised costs.[9] In other words, to improve costs, one must improve the total costs of health care from beginning to end.

As we show in greater detail, the same is unfortunately true of education. Improving the parts does not improve the whole system. Merely improving the quality of teacher education, having the best teachers in low-performing schools, giving students access to the latest and best computers, etc. do not by themselves necessarily improve the scores of students on standardized tests, let alone their ability to think creatively. It's not that these don't have an important effect. They do. But by definition, a system's performance depends upon the combined, interactive performance of the total system as a system.

Lastly, *a system has defining properties that none of its parts has.* Thus, purposeful motion is a property of the combined (i.e., interactive) human-machine system that is a car, and not just a vehicle alone. It is certainly not solely a function of the engine or wheels considered alone. Once again, without a driver or human interaction of some kind, e.g., remote control, a car cannot exhibit purposeful motion. Similarly, no

DOI: 10.1057/9781137386045

amount of analysis of the parts would possibly reveal a car's property as a social status symbol.

In the same way, "education" is a property of the whole system and not any one of its parts, no matter how good the individual parts are. This is not to deny in the least the tremendous effect that a single great teacher can have. One is affected by innumerable other equally important factors.

Problems versus exercises

The concept of systems has extremely important consequences for problems and especially for what counts as solutions.

Textbooks were, and still are, one of the prime pedagogic devices of The Machine Age. (The Machine Age is The Age of The Industrial Revolution. In The Industrial Revolution, all of nature was viewed as a machine that could be disassembled into its independent atoms, components, or parts.[10] Furthermore, the parts were completely interchangeable. If one part was defective, in principle it could be isolated, removed, and replaced with an identical part without affecting the performance of the whole machine. In contrast, in The Systems Age everything not only interacts with everything else, but stronger still, everything is integrally related to and is thus a part of everything else.) In turn, exercises are one of the prime features of textbooks. For instance, "If 400–X = 20, then find X," is an exercise. It is not a problem.

Exercises and problems differ in virtually every respect. Exercises are completely well structured and bounded. They are strongly differentiated and separated from everything that is not relevant to or a part of the exercise. Everything about them is known and defined precisely. This is exactly why they are artificial.

First of all, the complete statement of the exercise—not the problem—is *given* to the student so that there is no ambiguity whatsoever as to what is expected of him or her. In the simple example above, the student is expected to find the single number X *given* its precise relationship to the other numbers in the initial statement of the exercise. Second, there is generally one and only one right answer to every exercise. In the example, algebra and the laws of arithmetic guarantee that the answer is 380.

Problems have none of these characteristics. For one, context is everything. That is, *problems are part of systems*. They do not exist completely

DOI: 10.1057/9781137386045

on their own disconnected from a larger context, i.e., a system. Stronger still, *problems are abstractions from messes, which are even more complex systems.* In slightly different terms, problems are carved out of messes. In other words, problems do not have an independent existence of their own apart from a purposeful individual or individuals who have the problem.

Thus, if Sandra is a single mother with two kids to feed and has only $400 left at the end of the month, but needs at least $20 to pay for medicine for one of her sick children, then how much money does she have to spend for food and rent? What now is the *solution* to the problem? It is not just the simple number 380. Indeed, if Sandra really needs $500 to take proper care of herself and her children, what now is the *definition* of the problem? One of the places to look for a definition, let alone a solution, is in some other field of knowledge or profession such as family assistance and counseling, not just in algebra. If anything, the mere fact of the number 380 is more likely to lead to Sandra's frustration and even depression than to any sense of her having solved the problem.

As opposed to exercises, there are likely to be as many different formulations (definitions) of the problem and potential solutions as there are different stakeholders, i.e., all those parties that are affected by the problem, for instance, Sandra's parents, siblings, relatives, welfare agencies, and potential employers. Why should we expect everyone to have the same initial formulation of this or any problem? We shouldn't!

For this reason, problem negotiation is an important part of problem solving. In fact, the initial definition or formulation of a problem is one of the most important factors in its solution. But since exercises are preformulated, they generally do not teach students how to grabble with real, complex problems. If anything, they generally turn students—and even more unfortunately, adults—into "certainty junkies" so that if something is not well structured and predefined for them, then they experience noticeable discomfort and complain to the teacher or instructor.

Messes are even more complex and hence raise the ante even more.

The disposition of problems

There is another important aspect of problems we need to discuss. Ackoff makes a critical distinction between how problems are to be handled. There are at least four ways in which any problem can be managed.

DOI: 10.1057/9781137386045

Whereas an exercise, like a puzzle, can only be solved, and generally has one and only one solution, in contrast problems can either be *absolved, dissolved, resolved, or solved.* Furthermore, depending on the particular problem, its history, its current state, etc., all of these ways can be used at different times. They are not necessarily exclusive although they can be depending upon the particular problem and its context.

When we "absolve" a problem, in effect, we leave it alone in the hope that it will "right itself" or "just go away on its own." In many cases, this is nothing more than a case of wishful thinking. It can also be a case of outright denial. For instance, the unemployment problem (mess) certainly doesn't show any signs of going away on its own. Nonetheless, in some cases, doing nothing or not intervening, as in the case of Iraq, is the appropriate thing to do.

Alternatively, absolving can be a case of where we select a particular problem that we wish to focus all of our attention on, and thus divert our attention from others, even though the particular problem we have selected may not be the most critical one on which to work.

When we "dissolve" a problem, we attempt to redesign the underlying system or systems that gave rise to the problem in the first place. Or, we say that some other problem in the mess is more important and thus deserving of our attention. Therefore, we shift our attention, *but only after looking at the whole system or mess.*

When we "resolve" a problem, we accept a less than perfect state of affairs. For instance, economists typically "accept" an unemployment rate of 4 to 6% as "normal and/or acceptable." Most economists believe that attempts to go below these numbers would actually make things worse. For instance, to get full employment, wages might have to be lowered to unacceptable rates.

Lastly, when we "solve" a problem, we attempt to find the best or optimal solution to a problem. For instance, we attempt to make unemployment exactly equal to zero if indeed this is truly "best."

It should be clear from our earlier discussion that single, perfect, and exact solutions rarely exist for complex systems. In fact, so-called optimal solutions can make things worse as in the case of a car where a bigger and better engine without redesigning the entire suspension system can literally "backfire." Or in the case of health care where lowering the time that physicians spend with individual patients may save money in the short run but can actually lead to worse problems later on and thus raise costs appreciably.

DOI: 10.1057/9781137386045

The important point is that in managing messes, the best we can hope for is to dissolve them. Next best is to resolve them.

The notion that problems can rarely be solved is so important that we cannot stress it enough. This is in fact one of the cornerstones of psychodynamics:

> Freud...viewed suffering as inherent in the human condition and conflict as not only inescapable, but as basically irresolvable. To his mind, compromise solutions are all that can ever be had...For [the psychoanalyst] Mann, the ability to accept limitation and disappointment is the hallmark of the mature person. The hope that all problems can be solved and all goals achieved is what puts people at loggerheads with existence.[11]

Messes

We turn now to the definition of a mess.

A *mess* is a *whole system of problems* that is poorly organized, even disorganized. In fact, some of the disorganization is both intentional and *un*intentional.

None of the problems that constitute a mess even *exists*, and hence cannot be defined, independently of all of the other problems that are integral parts of the mess. A mess also contains the various parties—stakeholders in general—that play a major hand in defining the mess and who are affected by it. All the parties—stakeholders—who affect and are affected by a mess are *NOT* independent of it.

Since a mess contains stakeholders, it automatically contains all of their underlying anxieties, dreams, emotions, fears, hopes, and accompanying assumptions, beliefs, and myths, both conscious and unconscious. Furthermore, it contains as well the history associated with the mess, and potentially all other messes. As part of its history, it contains both strong conscious and unconscious memories of previous attempts, successful and otherwise, to manage the mess.[12]

In short, messes *potentially* contain everything pertaining to the human condition. (Even though a mess is not completely synonymous with the concept of "culture," a mess certainly contains the culture in which it is embedded and thus embeds.[13]) This is precisely why they are messy.

Similar to the definition of a system, *a mess consists of at least two different problems*. If there is only one problem, then it is neither a system nor a mess.

DOI: 10.1057/9781137386045

But, something else even stronger holds as well. *At least one of the problems in every mess is one of the problems from at least one other mess. The same holds true of assumptions, beliefs, myths, and stakeholders, etc.* Every mess thus contains at least one assumption, belief, etc. from at least one other mess. In addition, every mess contains at least one powerful underlying emotion from at least one other mess. In this way, every mess is in principle related to and a part of every other mess. *Thus, TEM is not independent of The Financial Mess, The Energy Mess, The Crime Mess, The Housing Mess, The Unemployment Mess, and so on.*

As an example, consider the following. It shows how low-income is inextricably intertwined with a host of social problems that affect the inability of low-income children to learn:

> Low family income makes it more difficult for parents to gain access to the high-quality child care that prepares children for kindergarten. It can also lead to classrooms filled with low-achieving, inattentive classmates. Crime in low-income neighborhoods may provide tempting alternatives to working hard at school and at the same time make it more difficult for neighborhood schools to recruit high-quality teachers. Plant closings can disrupt family life for children whose parents lose jobs, as well as deplete community resources that might have been channeled into school improvements.
>
> Rising inequality can have political repercussions as well. As the rich become increasingly isolated in certain neighborhoods and schools, the extent of inequality becomes less visible to them and to society as a whole, which can lead to increased social conflict and a reduced sense of common purpose. This can make it harder to mobilize the public concern necessary to deal with problems of disadvantage among those most at risk. Indeed, growing inequality can create a vicious circle: increasing returns to education create growing social and economic inequalities; these in turn exacerbate educational inequality and limit educational achievement among disadvantaged populations. Social and economic inequalities become more entrenched and limit social mobility, as more disadvantaged groups fall further behind ...[14]

For another, consider the following:

> ... Why is K-12 public school education a national security issue?
>
> First, it is critical that children in the United States be prepared for futures in a globalized world. They must master essential reading, writing, math, and science skills, acquire foreign languages, learn about the world, and—importantly—understand America's core institutions and values in order to be engaged in the community and in the international world.

DOI: 10.1057/9781137386045

> Second, the United States must produce enough citizens with critical skills to fill the ranks of the Foreign Service, the intelligence community, and the armed services. For the United States to maintain its military and its diplomatic leadership role, it needs highly qualified and capable men and women to conduct its foreign affairs.
>
> Third, the state of America's education system has consequences for economic competitiveness and innovation. No country in the twenty-first century can be truly secure by military might alone...[15]

A mess is also similar to a system in that no subset has an independent effect on the whole mess. In addition, a mess as a whole has properties that none of the "individual elements" have.

A mess raises the stakes even higher. It is a system but at a higher level of complexity. In brief, it is a "messier system."

Messes are also like fractals. The deeper and the further one digs down, one still encounters messes. It is messes all the way down and all the way up. Messes do not begin and end at any particular level of "reality." "Messiness" is an inherent property of all messes.[16] Thus, one does not reach any level of education where it is not messy.

Depending upon the number and types of wicked problems they contain—and each mess contains at least one *wicked problem—there are different types of messes. (A wicked problem is a problem that cannot be completely defined, let alone "solved," by any known, i.e., currently existing, academic discipline or profession either solely by itself or in combination with the others. Further, wicked problems do not have clear "stopping rules." That is, it's never fully clear when a wicked problems has been "tamed" and thus "solved.") A "wicked mess" is where "all or nearly all" of its problems are wicked.*

Messes are thus differentiated in terms of their "degrees of wickedness."[17] *And yet, because of their extreme importance, one cannot wait for new disciplines and professions to be invented before one copes with wicked messes or any mess for that matter. One must in other words manage (cope with) messes, wicked or not, in some way, where "copes with" means "to resolve," not "solve" them.*

Thus, TEM is not only a mess, but a prime issue is its degree of wickedness! It is certainly not a well-structured problem.

Finally, there are no neutral terms in a mess. While not all of them may be emotionally or value loaded to the same degree, ideology plays a central role. For example:

> ...it is nearly impossible to define harm—or danger, threat, menace—in a neutral way. Every definition of harm and its national security cognate rests on ideological assumptions about human nature, morality, and the good

DOI: 10.1057/9781137386045

> life. And in this regard, liberals are as guilty as conservatives. The only difference is that they often have less power to act on their convictions—and to stop their opponents from acting on theirs.[18]

Despite this, far too many academics and liberals still believe that "truth"—knowledge in general—and ethics are separable.

We build on and elaborate the preceding definitions of systems messes throughout the rest of this book.

Concluding remarks: the charter school mess

Figure 2.1 is a visual representation of The Charter School Mess. At this point, its primary intent is to show that none of its so-called "parts" even exists independently of one another. Therefore, none of the "parts" can be understood, let alone "coped with" or "managed," independently of one another. For the sake of simplicity, we do not draw arrows representing all of the possible interactions between all of the "parts," for to do so would overwhelm the reader, but to be sure, they are there. In principle, the arrows go simultaneously from every box to every other one, and back.

Naturally, we elaborate on Figure 2.1 in later chapters.

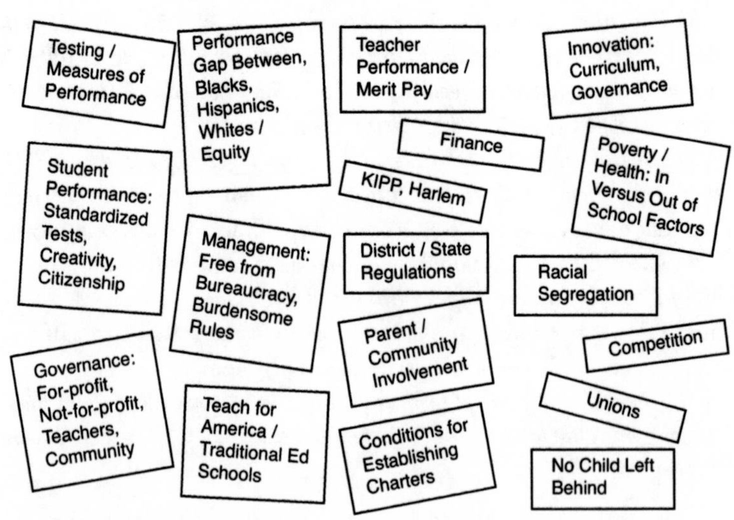

FIGURE 2.1 *The Education Mess—Charters*

DOI: 10.1057/9781137386045

To summarize briefly, *a "mess" is a whole system of problems that are so strongly interconnected such that they can't even be stated (formulated) properly, let alone dissolved and/or resolved, independently of one another. Strongest of all, none of the problems even exists independently of all the other problems that constitute a mess. To take any problem out of a mess is not only to distort the nature of the problem, but the entire mess. To "manage a mess" is to "manage its 'critical interactions'." But, what's "critical" can only be ascertained by looking at the entire mess.*

Lastly, we hypothesize that the overall measure of performance of a system is a *function* of the *total product* of the performances of all its interactions. It is definitively *not* the *sum* of the performances of its separate parts, for if it were, the parts would then be independent which they are not by virtue of being members of a system or mess. This means that if the performance of any one of the crucial interactions is zero, then one does not make up for it by improving the performance of any of the other key interactions. In short, a score of zero times a hundred is still zero. This notion holds as well, if not even more, for messes.

A critical point is that no one knows for sure, certainly not at this juncture, what the exact measure of performance of a mess is. We believe that for the reasons we have just elaborated in the preceding paragraph that whatever it is, the measure is a strong function of the interaction of the performances of the entire individual "parts." The simplest function for representing this is the multiplication of all the separate measures of performance. This does not of course make this measure "correct" or the only possible one. It is merely the simplest. However, a more important point is that *the measure of performance of a mess is itself one of the most important components of every mess!* This is certainly the case given that the measure of performance of any individual "part" is itself a function of the performances of all the other parts. Messes are complex entities indeed!

In Chapter 8, we demonstrate some key heuristics for coping with messes and demonstrate as best we can how they apply to TEM. We apply some of these heuristics in detail in Chapter 9 where we examine school safety and security as a mess. The important thing to note is that one only gets heuristics in dealing with messes. By definition, there are no ironclad rules that guarantee "exact solutions" to messes. If there were, we would not be dealing with messes.

Although we use and elaborate on these ideas and concepts throughout this book, and especially how they impact education, it is important to note that of course many analysts and interested observers recognize

DOI: 10.1057/9781137386045

the messiness of education. Our claim is not that we are the first or only to talk about TEM as a mess. For instance, many acknowledge that what a child brings with him or her to school, e.g., poor health, nutrition, poverty, etc., all greatly affect a child's ability and/or willingness to learn. But having said this, nearly all quickly fall back on one or two favored variables or remedies. In other words, they quickly abandon or shy away from the "messiness of the mess."

Once again, while we have no doubt whatsoever that good teachers can greatly affect a child's performance, we reject strongly the notion that the total performance of a system is solely dependent on this factor or any limited set of factors or variables. If it were, then it wouldn't be a mess, let alone a system.

We suspect that many, if not most, people shy away from messes by the fact that their education did not introduce them to the concept, let alone prepare them intellectually and emotionally for how to deal with them. They also have not been rewarded for it in their careers. But, there is also no question that certain personality types or dispositions experience considerable trouble in dealing with messes. Since systems can be said to have "dominant personalities," this means that certain fields and professions also experience great difficulties in handling messes. For this and other reasons that will become apparent shortly, we turn in the next chapter to the important role that personality factors play in systems and messes.

Notes

1 Quoted in Tough, Paul, *Whatever It Takes: Geoffrey Canada's Quest to Change Harlem and America*, Mariner, New York, 2008, p. 265.
2 Waddock, Sandra, *Not By Schools Alone, Sharing Responsibility for America's Education Reform*, Praeger, Westport, CN, 1995, pp. 23–24.
3 For a treatment of educational systems that comes close to this book, see Datnow, Amanda, *Integrating Educational Systems for Successful Reform in Diverse Contexts*, Cambridge University Press, New York, 2006.
4 Waddock *op. cit.*
5 Duke, Daniel L., "Tinkering and Turnarounds: Understanding the Contemporary Campaign to Improve Low-Performing Schools," in Stuit, David, and Stringfield, Sam, (eds), "Special Issue: Responding to the Chronic Crisis in Education: The Evolution of the School Turnaround Mandate," *Journal of Education of Students Placed at Risk*, Vol. 17, Nos 1–2, January–June, 2012, pp. 9–24.
6 Duke, *op. cit.*, p. 14.

DOI: 10.1057/9781137386045

7 Ackoff, *Re-Creating the Corporation*, *op. cit.*; Ackoff, Russell L. and Rodin, Sheldon, *Redesigning Society*, Stanford University Press, 2003; Ackoff, Russell L. and Greenberg, Daniel, *Turning Learning Right Side Up, Putting Education Back On Track*, University of Pennsylvania Press, Philadelphia, 2003; Gharajedaghi Jamshid, *Systems Thinking, Managing Chaos and Complexity*, BH, Elsevier, Boston, 2006; Gharajedaghi Jamshid, *A Prologue to National Development Planning*, Greenwood Press, New York, 1986.

8 Stuit, David, and Stringfield, Sam (eds), "Special Issue: Responding to the Chronic Crisis in Education: The Evolution of the School Turnaround Mandate," *Journal of Education of Students Placed at Risk*, Vol. 17, Nos 1–2, January–June, 2012.

9 Mitroff, Ian I., and Silvers, Abe, *Dirty Rotten Strategies: How We Trick Ourselves and Others into Solving the Wrong Problems Precisely*, Stanford University Press, Palo Alto, CA, 2009.

10 See Dolnick, Edward, *The Clockwork Universe: Isaac Newton, the Royal Society, and the Birth of the Modern World*, Harper, New York, 2011.

11 Alon, Nahi, and Omer, Haim, *The Psychology of Demonization: Promoting Acceptance and Reducing Conflict*, Routledge, New York, 2006, p. 30.

12 If there be any doubts about how central beliefs are to messes, then see Lawrence Lessig's powerful book. The beliefs of Americans that unbridled money corrupts the U.S. political system deeply affects their participation in the system: Lessig, Lawrence, *Republic Lost*, Twelve, New York, 2011

13 Smith, E.O., *When Culture and Biology Collide*, Rutgers University Press, New Brunswick, NJ, 2002.

14 Duncan, Greg J. and Murnane, Richard J. (eds), *Whither Opportunity: Rising Inequality, Schools, and Children's Life Chances*, Russell Sage Foundation, New York, 2011, p. 8.

15 Klein, Joel, and Rice, Condoleeza, Chairs, "U.S. Educational Reform and National Security," *Independent Task Force Report No. 68*, Council on Foreign Relations, New York, p. viii.

16 Sorensen, Geog, *A Liberal World Order in Crisis*, Cornell University Press, 2011; even though Sorensen does not use the term "mess" to describe the system of international politics, security, trade, etc., he is referring to it nonetheless; the international system is a mess at every level and type of activity, i.e., economic, ideological, institutional, structural, values, etc.; it certainly is a highly complex system in every respect.

17 The fear of course is that some problems such as global warming have become so critical that they are "super wicked problems." See Kelly Levin, Benjamin Cashore, Steven Bernstein, and Graeme Auld. "Overcoming the tragedy of super wicked problems: constraining our future selves to ameliorate global climate change," *Policy Science*, Vol. 45, 2012, pp. 123–152.

18 Robin, Corey, *The Reactionary Mind*, Oxford University Press, New York, 2011, p. 215.

DOI: 10.1057/9781137386045

3

The Psychology and Philosophy of Inquiry, Philosophical Psychology, and Psychological Philosophy

Abstract: *This chapter shows how the pioneering ideas of Carl Jung lead to a very different form of systems thinking, one that is especially suited to the analysis of education. It also introduces the concept of Inquiry Systems (ISs), i.e., different systems for producing knowledge. The chapter shows that for the most part the field of education has relied primarily on the wrong forms of inquiry. In short, education is an ill-structured mess, not a well-structured exercise.*

Mitroff, Ian I., Hill, Lindan B., and Alpaslan, Can M. *Rethinking the Education Mess: A Systems Approach to Education Reform.* New York: Palgrave Macmillan, 2013. DOI: 10.1057/9781137386045.

"Champions of solutions maintain that everyone knows what the problem is; but, in fact, most people know only their own experience with a given problem. It is not uncommon for citizens to spend their energy debating which of a number of predetermined solutions is best, seemingly unaware that there is no agreement on the nature of the problem..."

David Mathews[1]

"In an effort to clear up confusion (or ignorance) about the meaning of a word, does anyone ask, What is *a* definition of this word? Just about always, the way of putting the question is, 'What is *the* definition of this word?' The difference between *a* and *the* in this context is vast, and I have no choice but to blame the schools for the mischief created by an inadequate understanding of what a definition is. From the earliest grades through graduate school, students are given definitions and, with few exceptions, are not told whose definitions they are, for what purposes they were invented, and what alternative definitions might serve equally well. The result is that students come to believe that definitions were *not* invented; that they are not even human creations; that, in fact, they are... part of the natural world, like clouds, trees, and stars."[2] [italics in original]

Neil Postman

Introduction

In this chapter, we continue discussion of the basic concepts and ideas that are needed merely to think about TEM, let alone respond appropriately to it.

The first idea we introduce is the Jungian personality typology, and how it can be used to give deeper insight into the underlying psychological dimensions of complex problems. Different people not only instinctively prefer very different kinds of "solutions" to complex problems, but they "see" very different problems to begin with. In other words, they formulate very different problems from the very beginning of an inquiry.

One of the important implications of the previous chapter is that "problem formulation" is one of *the most critical* steps in problem solving. Indeed, many battles are fought over the "correct" definition of the

DOI: 10.1057/9781137386045

problem. Although the battles are generally not framed as such, they are really about the Type Three Error or E_3. The Type Three Error is defined as "solving the 'wrong' problem precisely."[3]

What good does it do to solve the "wrong" problem precisely? Not much! And in fact, it's worse than producing the wrong solution to the right problem. The right solution to the wrong problem keeps one stuck in going down the same wrong path whereas ideally the wrong solution to the right problem allows one to learn from one's mistakes.

The second idea is that of Inquiry Systems, i.e., which types of inquiry or knowledge systems are "best-suited" for which kinds of problems?[4] Ill-structured problems are not the same as well-structured problems or exercises, and hence, cannot be treated in the same way.

Despite all the lip service to the contrary, education has still mainly been treated as if it were a well-structured issue or problem when it is highly ill-structured and messy. This is true even of those aspects that are comparatively well structured such as the design, use, and analysis of standardized tests. Since tests only exist and function within a larger mess, they are far from being entirely well structured, certainly not in their interpretations and implications.

The concepts of this and the previous chapter are necessary to understand why education is not only a complex system, but a mess. Unless one understands the true nature of education in the 21st century, then one not only perpetuates the mess, but makes it worse.

The ideal organization

In the early 1970s, Mitroff and a life-long friend and colleague Ralph Kilmann started their academic careers at the Graduate School of Business at the University of Pittsburgh. As a result of their working closely together, they made a basic discovery about the Jungian personality types. The discovery was so impactful that it forever changed their thinking, if not their entire careers.

For those who are not familiar with the Jungian personality framework, we describe it shortly. More importantly, we show how it can be used as a group problem-solving methodology, and ultimately as a framework for systems' thinking. In other words, the Jungian personality framework is not confined to the study of individuals alone. The Jungian personality framework may have originated as a framework to

DOI: 10.1057/9781137386045

understand individual personality differences, but it is not confined to the study of individuals. It can be used to illuminate the nature of complex problems. As a matter of fact, it is particularly suited to the study of complex problems and messes such as education.

This does not mean that the Jungian personality framework is the only one that can be used for this purpose. It is merely one that the authors have found that is very fruitful. We thus use the Jungian personality framework primarily as an interpretative framework, not as a personality instrument.[5]

The Jungian personality framework is based on the path-breaking work of the Swiss psychiatrist and psychoanalyst Carl Jung.[6] As was the case with virtually all the well-educated people of his time, Jung was extremely well versed in European culture, history, literature, and philosophy. As a result, Jung noticed that the same psychological differences or types emerged between different artists, authors, analysts, poets, philosophers, etc. repeatedly no matter what the particular subject matter. Jung codified the underlying differences into a system, which later served as the basis for his personality framework.

The context for the discovery was that Mitroff and Kilmann had to conduct a workshop for teachers about different kinds of organizations. They felt an acute sense of anxiety because they literally didn't know what they were going to do. They only knew that of all parties, they didn't want to lecture teachers for eight hours straight.

Since they both knew the Jungian personality framework, an idea literally popped into both of their heads instantaneously. Suppose one put all the individuals with the same Jungian personality type into a single group. In this way, there would be different groups for each of the separate personality types. Furthermore, suppose they gave each group the same assignment to see whether one could bring out the differences between them in such a way that everyone could clearly see the effects of personality. To help make this possible, the assignment was not only for each group to *describe* their "Ideal Organization" and give it an identifying name, but to *build* something out of, say, tinker-toys that would illustrate it. If it were successful, it would externalize an internal predisposition that by definition is difficult, if not virtually impossible, to see.

The assignment to "describe your ideal organizations" was purposefully left ill-defined so that each group would project their personality onto it. Giving the groups a detailed, well-defined exercise would in effect be imposing the demands of one personality type on all of them.

DOI: 10.1057/9781137386045

Mitroff and Kilmann were taking a big gamble because they didn't know whether the *behavioral exercise* they created would work or not. That is, the assignment was certainly an exercise all right, but it was not in the sense of a simple, predefined exercise in, say, algebra with a single "right numerical or symbolic answer." And, if it didn't work, they were in trouble because the only backup they had was to lecture about organizations.

Fortunately, the experiential exercise not only worked, but it worked exceptionally well. Mitroff and Kilmann had not only stumbled onto Jungian group, or social psychology, but more importantly, they had found a way to use the Jungian personality framework for group problem solving.

As we shall see, the Jungian personality framework is extremely helpful in understanding how and why different people and groups formulate TEM differently.

The Jungian personality framework

The Jungian personality framework contains four separate dimensions. It also contains two opposite personality functions (or types) that bind each end of the separate dimensions: (1) Introversion or I versus Extroversion or E; (2) Sensing S versus Intuition N; (3) Thinking T versus Feeling F; and (4) Perceiving P versus Judging J. Since there are two possible choices or outcomes for each dimension, there are 2 times 2 times 2 times 2 or 16 different personality types that can be formed by combining the ends of each of the four dimensions in all ways, e.g. ISTP, ESFJ, etc. Since 16 groups are a lot to work with, Mitroff and Kilmann chose to work with only four. These four are captured succinctly in the four domains or quadrants in Figure 3.1.

> Domain 1—the upper left quadrant—represents an impersonal, technical, scientific approach to the study of parts (ST);
>
> Domain 2—the upper right quadrant—represents an impersonal, technical, scientific approach to the study of wholes, i.e., systems (NT);
>
> Domain 3—the lower right quadrant—represents a personal, social, cultural approach to the study of groups and/or whole communities (NF);
>
> Domain 4—the lower left quadrant—represents a personal, social, cultural approach to the study of particular individuals (SF).

DOI: 10.1057/9781137386045

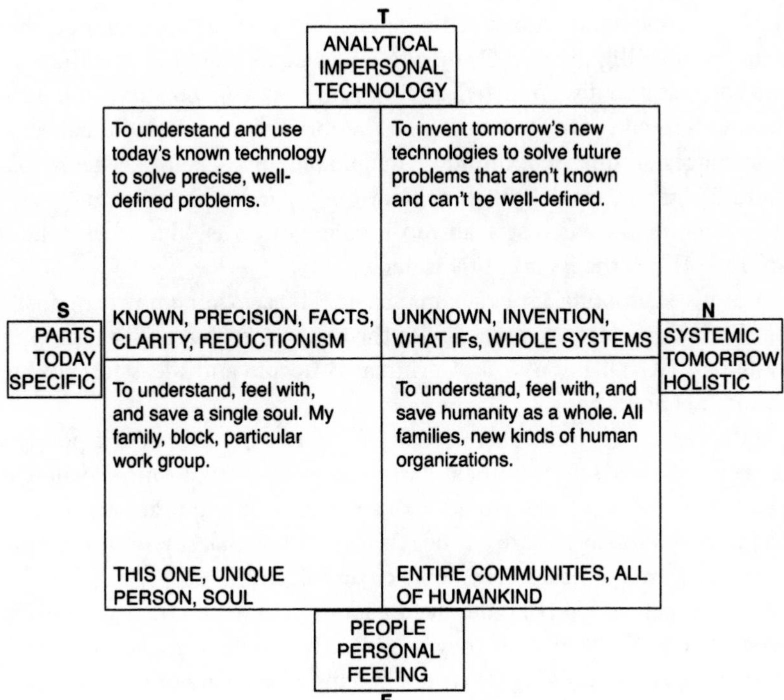

FIGURE 3.1 *The Jungian framework*

Sensing-Thinking Types (Domain 1 or STs for short) instinctively break all problems and situations down into independent, detailed parts for which they gather "hard and precise data." This is the S part of their personality. The parts and the data are then analyzed impersonally according to accepted modes of thinking, i.e., conventional logic and the well-established principles of contemporary science. This is the T part.

Intuitive-Feeling Types (Domain 3 or NFs) are the complete opposite of STs. NFs not only look at the big picture in the form of whole communities, but they look at everything in intensely personal and human terms. (Once again, N stands for Intuition; F for Feeling.) Where only details and thinking matter to STs, only the big picture and feelings matter for NFs.

Sensing-Feeling Types (Domain 4 or SFs) are focused on particular individuals. They break complex systems and problems down to the personal impacts they have on specific individuals. In other words, their

DOI: 10.1057/9781137386045

S part is concerned with specific people, not with detailed, impersonal data or facts. Big pictures (N) make no sense to them because they are too abstract, too disconnected, and far removed from concrete individuals and human feelings. Whether in the form of the whole human race (the biggest human community imaginable) or gigantic systems, big pictures literally don't exist for them. Only particular individuals with their unique idiosyncrasies, all-too-human concerns, likes, and dislikes are real. This is the F part of their personality.

Intuitive-Thinking Types (Domain 2 or NTs) are the complete opposite of SFs. NTs are also concerned with the big picture, but only as complex systems that call for new and original concepts and ideas for how to think (T) about them.

Introversion (I) and extroversion (E) refer to the source of a person's energy. Introverts derive their basic life energy from within whereas extroverts derive it from groups. Contrary to common usage and misinterpretation, Introverts are not necessarily anti-social, shy, or quiet. And, Extroverts are not necessarily noisy or unreflective.

Finally, perceiving (P) types need a lot of hard data if they are sensing types, and a lot of ideas if they are intuitive types to make or reach a decision whereas Judging (J) types need little hard data or ideas to make an important decision or to reach closure.

Four ideal organizations

In order to more easily understand the Jungian personality framework, we have deliberately exaggerated the differences between the types in what follows.

Most people fortunately are a mixture of the various Jungian types so that they can communicate and get along with one another. Nonetheless, the extremes do exist. More importantly, the extremes are helpful in understanding the Jungian personality framework. Without the extremes, one easily gets lost in inevitable misunderstandings and overlaps between the types.

To return to the experiential exercise, the ideal organization of Sensing-Thinking types (STs) is bureaucracy. If bureaucracy had never existed, it wouldn't matter because STs instinctively recreate it.

STs love bureaucracy because everyone knows precisely what is demanded of them and how they will be evaluated. Furthermore,

DOI: 10.1057/9781137386045

everyone's job is specified far in advance and excruciating detail. The jobs are also fixed; they never vary from day to day. In addition, there is a small set of "outcome measures" on which everyone and the entire organization is evaluated and held strictly accountable. Typically it is "contribution in dollars to the 'bottom-line.'" There are other measures as well, for instance, how many "widgets" a person or department produces, amount of waste, lost hours—in short, anything that can be measured precisely. If it can't be measured, then not only does it have no meaning, but it literally doesn't exist for STs. In short, there is no ambiguity whatsoever in the ideal world of STs because they despise ambiguity with a passion.

(The fact that every one of the types can be highly emotional in favor of their preferred view of reality—indeed, what they regard as reality—shows that Feeling F and emotions are not the same. Feeling is a way of apprehending and responding to the world. It is not necessarily emotional per se.)

In this way, there is no ambiguity whatsoever. Everything is well-defined and in its "proper place." It comes as no surprise that more often than not the thing that STs build out of Tinker Toys to represent their ideal organization is a symmetrical, well-ordered, highly efficient, energy machine of some sort.

In sharp contrast, Intuitive Feeling types (NFs) literally hate bureaucracy because it "hems them in." It constrains human imagination and feeling. It treats all people as if they were merely interchangeable parts of some gigantic, impersonal machine. NFs typically build something that not only represents all of humanity, but is "free floating and unconstrained." As a result, it is highly asymmetrical. Asymmetry stands for the strong belief of NFs that "humans cannot be put into narrow, prearranged boxes."

In a recent workshop that Mitroff conducted with educators, the members of the NF group formed themselves into a circle to symbolize their "interconnectedness." Each individual then held up a different picture of a person to indicate his or her concern with all of humanity.

NFs not only prefer organizations that are completely "flat," i.e., without any fixed or permanent hierarchy, but in the extreme, no structure whatsoever. NF organizations—"random groupings" is a better term—do not have permanent, fixed job descriptions, leaders, or "well-defined measures of performance" that are specified in exacting detail. People just show up and somehow just "come together." How they happen to

DOI: 10.1057/9781137386045

feel on a particular day determines what and how they do, and who gets together with whom.

To say that STs and NFs are the complete opposite of one another is putting it mildly. More often than not, they literally cannot understand, let alone stand one another. They don't know what they are talking about.

People mistakenly think that they *only* need translators for people from different countries—for example, China and France—who speak very different languages. But they need translators even more for people who seem to be but are not really speaking the same psychological language, e.g., different forms of English.

STs cannot understand how NF organizations can exist, let alone function without any rules or structure. And, NFs cannot understand how STs can live without the "joy" of constant uncertainty, human creativity, openness to change, and a basic concern for humanity, not money or widgets.

Intuitive Thinking (NT) organizations are best described as "matrix organizations." Neither the jobs nor the people in them are fixed. Depending upon the particular problem, particular task groups are formed. The groups have the variety of skills necessary to solve a particular problem. As new problems arise, the groups vary in composition and skill level. And, so do the measures of performance required to specify when a particular problem is "solved."

NT organizations value imagination, novelty, change, unpredictability, and a focus on the "whole system" and the "big picture." They especially love inventing new ideas and concepts and applying them to new scientific and technical problems. Thus, NTs build something that represent new systems for "putting it all together."

Sensing Feeling (SF) organizations are more akin to small families than anything else. SF organizations pride themselves on everyone knowing and caring about everyone else. They get together as often as they can. They have frequent family picnics. They know and celebrate one another's birthdays, anniversaries, etc.

SFs do not come to work primarily to make money, although they certainly need it as much as anybody. They primarily come to work to find meaning and purpose in small groups that really care about them. SFs build an image of a particular person.

Once again, it is putting it mildly to say that NTs and SFs don't understand one another. NTs don't understand the parochialism of SFs and their constant "gushiness" about feelings. And, SFs don't understand the abstractions of NTs that "rob them of their down-to-earth humanity."

DOI: 10.1057/9781137386045

STs and NTs share at least one psychological function in common, T, analytical thinking. And, NFs and SFs share a concern with people and feeling. But, STs and NFs share nothing in common. And, neither do NTs and SFs.

By describing their ideal organization, listing its chief properties and characteristics, giving it a name, and building a representation of it out of Tinker Toys, one is literally able to "see" an internal personality trait that is generally invisible to the ordinary eye. One is also able to take the Jungian framework out of its original context as an instrument for measuring individual personality traits and show how it applies to group problem solving. In this way, the Jungian framework not only helps to explain group behavior, but it can be used as a guide to produce at least four very different formulations of complex problems, i.e., problems which by their very nature require more than one definition so that we don't end up solving the "wrong problem precisely."

If one uses all of the Jungian types, then one can in principle produce 16 different representations of any and every problem.

The train crossing problem

Consider another behavioral exercise that we have used as well. It also happens to be a real problem.

It seems that no matter what the latest safeguards, each year children still get killed at dangerous train crossings. The Jungian framework provides an important way to analyze the variety of approaches that different psychological perspectives instinctively take in responding to a tragedy that need not and should not happen.

STs naturally see the problem as one of installing better warning and signaling devices both on-board trains and at dangerous crossings. Thus, far in advance, when a train approaches a crossing, a signal would sound repeated warnings to the engineer to slow down, "smart computer operated" gates to close on their own, loud horns to sound, etc. The trains might even slow themselves down automatically without any input from the engineers. All of this would be accomplished with today's proven technology, i.e., without inventing anything new.

NTs instinctively want to look at the whole system and systems in other industries, even other countries. For instance, is the pay incentive system so out of whack such that train engineers are rewarded for sticking to

DOI: 10.1057/9781137386045

their schedules, and even beating them, instead of ensuring safety? What good does it do to install the latest safety devices if engineers are not rewarded for using them? Furthermore, if one takes a broader look, what can be learned from other high-risk systems such as nuclear power plants and other train systems around the world? (Given the nuclear disaster in Japan in 2011, the question needs to be reversed: what can nuclear power plants learn from other critical systems that need to be operated safely?) If new technology needs to be invented, then so be it.

NFs take a completely different tack. They want to educate the children and the community to the inherent dangers so that together they can take collective action. For instance, independent of technology, at crucial times of the day, parents need to come together as a group to walk the children safely across the tracks. In this way, they not only solve an important—literally life and death—problem, but they show their direct concern for the children and the community as a whole.

Finally, SFs want to come to the sites of past tragedies and place pictures of specific children who have been killed, share cherished mementoes, light candles, read poems, and leave toys. They not only want to grieve with the children and their parents, but to heal their own souls that have been seriously wounded by the tragedies. This is precisely why one sees so often at great tragedies, e.g., 9/11, reams of poems, candles, small animals, and mementoes that are left to honor the dead.

One of the unfortunate, unanticipated side problems of the Sandy Hook Elementary School shooting in Newtown, Connecticut is that over 800 hundred items of small animals, clothes, poems, etc. have been donated by people all around the U.S. in sympathy with the victims. Despite the best of intentions, the overall effect has been to overwhelm the capacity of the volunteers to process the items. The point is that none of the Jungian types can manage the world without the explicit cooperation of all the others. In the case of Newtown, SF needs ST to help manage the process of dealing with the nation-wide outpouring of feeling.

SFs also want to have small groups walk the children safely across the tracks. In this case, they want families who live in close proximity on particular blocks to take turns walking the children. They want the children to see neighbors that they know and trust.

Every one of the types is solving a different problem because internally they see a different problem. STs see the problem strictly as one of applying known technology to a strictly well defined, precise, and bounded issue. However, the real problem they are attempting to solve is buried

DOI: 10.1057/9781137386045

deep in their unconscious. STs attempt to manage their own unacknowledged anxieties by constructing a bounded, well-ordered universe in which they feel completely safe and protected. But, since they generally put down emotions of any kind, they are often completely out of touch with their own feelings, let alone those of others. For this reason, they are often perceived as cold and uncaring.

NTs see the problem as one of the "whole system." By focusing on new technologies, they often avoid the here-and-now and dealing with their own feelings as well. They are often perceived as abstract, distant, and uncaring.

NFs not only see the problem in human terms, but as one that involves the whole community. While this is fine, they often do so because they have a deep-seated fear and distrust of technology. For this reason, NFs are often perceived as fuzzyheaded and soft.

Finally, SFs are solving the problem of dealing with their own grief. While this is fine and even commendable, they often do so because they also have a deep-seated fear and distrust of technology. More often than not, they are perceived as gushy and softheaded.

All four types are needed in responding to any problem of even moderate complexity. Ideally, all, four should work together and complement one another. Every problem has aspects that involve both conventional and novel technology. Every problem also has deep human and feeling dimensions. Indeed, without engaging people at their deepest levels, how can technology be effective?

In this sense, all of the Jungian types taken together constitute a human system. Far more than they realize, all of them are *inter*dependent, not *in*dependent. None of them can really exist, let alone function without the others. Without the support of a broader community, how could STs get up every day and function? Without rules and structures, how could NFs even exist and function as well, and so forth for the other Jungian types?[7]

An education example

Let us give a brief example of how the Jungian types easily enter into the field of education. Reflecting on the Turnaround Movement, Michael Hansen writes:

> In spite of the interest in intervening in the bottom 5% of the nation's chronically low-performing schools, the definitions of the key concepts remain

> nebulous. The labels *chronically low performing* and *turnaround* have become ill-defined buzzwords that are commonly used in policy discussions, bent and shaped to fit the purpose at hand. [italics in original] Although states, practitioners, and researchers have employed a variety of approaches in identifying schools to target, no standard exists on what qualifies a school to be eligible for these designations.... *this unsystematic approach to turnaround is dangerous.* [italics added][8]

The above is as perfect an example of a "pure ST statement" that one could ever hope to find. Notice carefully that we are not saying that Hansen himself is an ST for apart from directly administering a psychological test, we have no way of knowing his or anyone else's "type." Nonetheless, in the spirit of Jung, we do not have to know one's type in order to analyze statements for their underlying psychological orientation.

Consider another case, the Roxbury Preparatory Charter School. Merseth writes:

> ... planning efforts begin in August, when staff members work full-time for three weeks to prepare for the academic year. This preparation is comprehensive [thus, ST and/or NT depending upon the type of "comprehensiveness"]: it includes a discussion of long-term goals for the year [this is NT and/or NF depending upon the type of "discussion" and "long-term goals"]; the orientation of new staff members to schoolwide [sic] systems [NT] and routines [ST]; role-playing that depicts common classroom challenges [SF, NF]; and conversations about race, diversity, and the importance of communicating with families [SF, NF] ...
>
> ... In other words, the planning process ensures that teachers think strategically [NT] ...[9]

Notice carefully that in the above quote, all of the four Jungian types are represented depending upon the interpretation of the meaning and intent behind the words. That is, in many cases, one can definitively say with little doubt that a proposition or statement clearly represents a particular Jungian orientation. However, in other cases, we need more information before we can definitively classify a particular statement. Even more importantly, depending upon the psychological meaning behind it, any statement can be an example of any of the types.

Finally, the difference between different Jungian types is also one of epistemology, i.e., which kind of knowledge systems the different Jungian types instinctively prefer even though not all of them are appropriate for ill-defined problems. For this reason, we turn to Inquiry Systems.

DOI: 10.1057/9781137386045

Inquiry Systems (ISs)

ISs are not only systems for producing fundamentally different kinds of knowledge about a problem, but they also differ basically in what they regard as knowledge and how to obtain it.

ISs are tightly coordinated systems consisting of (1) inputs, (2) operators, (3) outputs, and (4) guarantors. The inputs are the basic building blocks or starting points of knowledge. The operators are the mechanisms that operate on the incoming inputs and transform them into valid outputs, i.e., knowledge.

For example, in the Hansen example in the preceding section, unless we start with the exact definitions of key terms (precise inputs), then we will never be able to arrive at valid conclusions (outputs)—in this case, turning around low-performing schools.

The guarantors are the most important, and thereby the most basic, parts of ISs. The guarantors exist for the basic purpose of assuring (guaranteeing) that if one starts with the right kinds of inputs, operates on them in the right ways, then one will obtain valid knowledge, i.e., "truth." In this sense, the perpetual battles in Western philosophy over what "truth" is and what is the best way of obtaining it might better be termed "guarantor wars!"

In effect, the guarantors are the "supreme controllers" of an IS for they specify the "right kinds" of inputs, operators, valid types of outputs, etc.

While there are many different types of ISs that in turn assume many different forms, we're going to concentrate mainly on just five. These are sufficient for our purpose.

They are (1) Expert Consensus ISs; (2) Scientific Modeling or Rationalist ISs; (3) Multiple Model ISs; (4) Expert Disagreement or Dialectical ISs and finally, (5) System Thinking ISs. These five are found repeatedly in the history of Western thought. In this sense, they are archetypal ways of knowing.

Expert Consensus (Empirical) ISs

Expert Consensus ISs are precisely what their name implies. They start with simple facts, observations, and/or the judgments of experts—expressed in the form of various types of numerical inputs—and then operate on them in such a way so as to reach empirical knowledge or

DOI: 10.1057/9781137386045

truth. The operators are typically various processes by which one obtains the average of a set of numbers. The guarantors are the "tightness of agreement" between the initial observations, etc. The "tightness of agreement" is typically measured by the spread or standard deviation between the initial numbers. Supposedly, the tighter (narrower) the spread, then the better (more certain, reliable, etc.) the resulting output or "truth" of the system.

Expert Consensus ISs are well suited for—indeed they are *only* well suited for—highly well-structured problems for which one can not only gather precise data, but for problems that can be broken down into parts such that precise data can be gathered on the parts. Furthermore, because they believe that there is one and only one best empirical explanation for all phenomena—one set of facts or numbers—Expert Consensus ISs tend to produce only a single set of resultant outputs. In this way, everything can be reduced to or summed up in a single number such as the average or median of a set of numbers. They also believe strongly that data are theory free, that one does not need to presuppose any theory in order to gather data.

It should come as no surprise that Expert Consensus ISs not only appeal primarily to STs, but are their creation. With their great emphasis on data, they certainly appeal to the S side of STs.

Scientific Modeling (Rationalist) ISs

Scientific Modeling ISs typically start with a set of basic ideas that are so "intuitively obvious" that no "rational person" could doubt their "truth." Euclidean geometry is the classic example. One starts with sets of basic intuitive notions about points, lines, and plane surfaces that are accepted as "given." The basic starting ideas are then combined through logical and mathematical operations so as to produce logical truths, i.e., theorems. The guarantor is typically the "law of contradiction" or the "law of the excluded middle," i.e., an assumption, idea, proposition, or a theorem cannot be both true and false at the same time. For example, in Euclidean geometry, a line A cannot be both perpendicular and parallel to another line B at the same time. It's one or the other, but not both.

Scientific Modeling ISs also typically build one model of any phenomenon for they have a fundamental assumption that "truth is singular." There is one best logical explanation or mathematical model for any

DOI: 10.1057/9781137386045

phenomenon. In this way, they generally presuppose one and only one scientific discipline. They also assume that data and theory are independent of one another in the sense that logic is presumably independent of facts.

Scientific Modeling ISs are also only well suited for extremely well-structured problems such that they can be reduced to simple propositions and/or variables and hence modeled.

It should come as no surprise that Scientific Modeling ISs not only appeal primarily to STs, but are their creation as well. With their great emphasis on abstract ideas, formulas, and models, they certainly appeal to the T side of STs.

Multiple Model ISs

Multiple Model ISs start with multiple models of any phenomenon. Multiple models are thus the inputs. They assume that reality is too complex to be captured in any single data set or model.

Ever since the great German philosopher Immanuel Kant, philosophers have recognized that all data are theory and value dependent. One cannot gather data independently of the presupposition of some theory. All data presuppose theories in the sense that theories guide us as to what is important to collect in order to understand a phenomenon. One must have some theoretical understanding of what the phenomenon is in order to gather relevant data about it. But different theories automatically lead to the collection of different data.

Consider drug dependency. Medical models of drug dependency are not the same as psychological, family systems, economic, etc. models. They each start with different basic notions and as a result collect different supporting data. In addition, it is definitively not the case that one model is "right" and all the others are "wrong." They are merely different because they start with different assumptions and "explanatory variables."

The operators of Multiple Model ISs are all-too-human beings that need to be well informed with regard to some phenomenon before they can make an intelligent decision with regard to it. "Being well-informed" means seeing how a phenomenon "changes" as we view it from multiple perspectives. Hopefully, in this way, the decision maker and his or her colleagues can integrate the various perspectives in their minds so as to

DOI: 10.1057/9781137386045

reach a "richer decision," i.e., the output. Of course, like all ISs, there is no absolute guarantee that the guarantor of being well informed will work. In this case, the guarantor is the process itself of witnessing the explicit operation of multiple models with respect to the "same phenomenon." At best, the guarantor is a minimal guarantee that one will not solve the "wrong" problem precisely. But, there is always the possibility that one will end up being and feeling more confused. For this very reason, not all ISs are equally well suited for all Jungian types.

Notice that this system is for moderately ill-structured problems that can be understood well enough such that they can be modeled from different perspectives. The prime purpose of looking at multiple representations of a phenomenon is because one is in doubt as to the "true or real" nature of the problem.

It should come as no surprise that Multiple Model ISs not only appeal primarily to NTs, but are their creation. With their great emphasis on multiple ideas, formulas, and models, they certainly appeal to the N and T sides of NTs.

Expert Disagreement ISs

Expert Disagreement ISs are the complete antithesis of Expert Agreement ISs. The prime assumption here is that a decision maker will learn more from witnessing a strong debate between two diametrically opposed positions on any issue than quick or facile agreement. This system is for highly ill-structured problems, i.e., problems for which there are strong disagreements as to the "nature of the problem." If all one had were Consensus ISs, then inquiry would be over before one even started because consensus is lacking from the very beginning!

President Franklin Delano Roosevelt (FDR) is a prime example of a chief executive who not only believed strongly in this particular IS, but used it regularly. This doesn't mean that FDR necessarily knew about ISs!

Whenever FDR had to make a critical decision, he assigned two different aides to research it from two diametrically opposed perspectives. Of course, beforehand, the two aides didn't know that FDR was doing this.

When they were ready to make their reports, FDR brought the aides back to report to him in person at the same time. While they were annoyed with having to make their reports in front of one another and

DOI: 10.1057/9781137386045

the fact that they didn't agree, FDR felt he was not informed unless he first heard the strongest possible debate on an important issue. He and his other aides then did the best they could to synthesize the two different views and hopefully get the best of each.

In this case, the inputs are the two very different views, and the operator is the debate itself and the process of synthesis, which is also the guarantor.

With their emphasis on debate and human disagreement, Expert Disagreement ISs appeal to NTs and NFs.

Systems Thinking ISs

Systems Thinking ISs are what this entire book is all about. Systems Thinking is not only about the definition of systems and messes that we described in the previous chapters, but Systems Thinking ISs are about expanding as much as possible our conceivable views of any and all phenomena. But, they are much more than this.

In Systems Thinking the epistemic (the true), the ethical (the good), and the aesthetic (the beautiful) are inseparable. Whether one is aware of it or not, every inquiry presupposes some theory of knowledge or truth, ethics, and aesthetics. Furthermore, they are highly interdependent. They not only strongly influence one another, but they are part of one another.

Systems Thinking ISs are the embodiment of the philosophical school of Pragmatism. A succinct phrasing of Pragmatism's definition of "truth" is: "*Truth is that which makes an ethical difference in the quality of one's life.*" Whatever one's prevailing definition, notion, and theory of ethics, in Pragmatism, truth and ethics are inseparable. Furthermore, the small word "makes" means we do not have truth unless something *makes* a profound difference in one's life, i.e., until it is implemented. Truth and implementation are not separable as well. Finally, "quality" is a stand-in for aesthetics. Thus, the true, the good, and the beautiful are inseparable in Pragmatism.

Systems Thinking ISs also make use of all the other ISs. In the beginning, problems need to be viewed from multiple perspectives and their nature debated. Only later when one is satisfied as to the underlying nature of the problem does one bring in empiricist and rationalist modes of inquiry, i.e., the first two ISs.

Another strong tenet of Systems Thinking is that *there are no basic disciplines* to which all others can somehow eventually be reduced. Every

DOI: 10.1057/9781137386045

one of the academic disciplines and professions are "basic" in the sense that they are all a valid component of human inquiry and systems. Each has a fundamental contribution to make to knowledge. Furthermore, all of them are interdependent, not independent. For instance, given that everything we know is done by and for humans, psychology is a part of every investigation and human action whether we acknowledge it or not. Lastly, since Systems Thinking ISs have an expansive view of "knowledge," they strive to incorporate all of the Jungian types.

SF Inquiry

The perceptive reader will undoubtedly have noticed that none of the Inquiry Systems we have discussed thus far appeals directly to and/or are based on the Jungian function SF. This is because the dominant Inquiry Systems we have discussed are the product of Western civilization. As such, primarily men developed them. As a result, they do not exhaust the full range of inquiry.

While this is not the place to discuss the full range of different forms of inquiry that are based on SF (they incorporate NF as well), we would be remiss if we didn't note that Feminist Inquiry largely arose out of the neglect of SF.[10]

In Feminist Inquiry, knowledge is fundamentally the product of deep, caring, sustained human relationships. In this sense, neither the Inputs, Operators, nor Guarantors of Feminist Inquiry are primarily cognitive. Above all, they are certainly not impersonal mechanisms. This does not mean that they are devoid of logic or reason. Rather, they are based on the "reason" and "logic" of caring relationships, i.e., Feeling in the Jungian sense.

All inquiry in the beginning and end is the product of human beings. Unless there is a "deep spirit of cooperation" between the people involved in inquiry, nothing fruitful will result.

Concluding remarks

In the rest of this book, we put the ideas of the first two chapters to use in understanding TCSM and TEM. Because the ideas are complex, we revisit them repeatedly. Our claim is that it is not that the ideas and concepts

DOI: 10.1057/9781137386045

of this chapter exhaust all the various forms of inquiry, but that they are necessary, to understand phenomena as complex as TCSM and TEM.

We need to make one thing as clear as possible. Expert Consensus (Empirical) ISs have been the primary basis for evaluating and critiquing the vast majority of educational efforts, e.g., how charter schools fare relative to public schools with regard to scores on standardized reading and math tests. That is, we use the averages of standardized test scores for various groups of students, schools, etc. as the primary operator and even guarantor to compare how students fare in different types of school environments.

While all efforts certainly need to be evaluated against the best data available—we have no objection to this as far as it goes—it needs to be kept firmly in mind that Expert Consensus (Empirical) ISs are only good for those aspects and types of problems (messes) that are well structured. We cannot emphasize enough that TEM is not a well-structured problem. Therefore, it cannot be treated as if it were. It is not even clear that ordinary test and other data are sufficient to evaluate something so complex as TEM. As we shall see, this has important consequences for types of data that are collected and how they are used to improve education, not to mention to evaluate the performance of individual students.

Finally, the Jungian framework, particularly in the ways that we use it as a group problem forming and solving method, is a mixture of various ISs. First, the Jungian framework is Multiple Model IS in that it generates at least four very different views of any problem. Second, it is an Expert Disagreement IS in that it produces strong conflict between different views. Third, it is a Systems IS in that it attempts to integrate the different views into a coherent whole such they support one another. In this way, the Jungian framework and ISs not only work together, but are integral parts of one another.

Notes

1 Mathew, David, *Is There a Public for Public Schools?*, Kettering Foundation Press, Dayton, OH, 1996, p. 48.

2 Postman, Neil, *The End of Education: Redefining the Value of School*, Vintage Books, New York, 1996, p. 172.

3 Mitroff, Ian I. and Silvers, Abraham, *Dirty Rotten Strategies: How We Trick Ourselves and Others into Solving the Wrong Problems Precisely*, Stanford University Press, Palo Alto, CA, 2011.

DOI: 10.1057/9781137386045

4 *Ibid*; see also Churchman, C. West, *The Design of Inquiring Systems: Basic Concepts of Systems and Organizations*, Basic Books, New York, 1971.
5 See Smith, Christian, *Moral Believing Animals*, Oxford University Press, New York, 2003, for a particularly illuminating discussion of interpretative frameworks.
6 Jung, Carl, *Psychological Types*, Vol. Six, *Collected Works*, Princeton University Press, 1971.
7 Indeed, countless works in contemporary psychology have shown that the traditional, strict separation between (1) cognition or reason and (2) emotion or feelings is a false and misleading dichotomy. For the latest, see Haidt, Jonathan, *The Righteous Mind, Why Good People Are Divided by Politics and Religion*, Pantheon, New York, 2012.
8 Hansen, Michael, "Key Issues in Empirically Identifying Chronically Low-Performing and Turnaround Schools," in Stuit, David, and Stringfield, Sam (eds), "Special Issue: Responding to the Chronic Crisis in Education: The Evolution of the School Turnaround Mandate," *Journal of Education of Students Placed at Risk,* Vol. 17, Nos 1–2, January–June, 2012, 55–56.
9 Merseth, Katherine, *Inside Urban Charter Schools: Promising Strategies in Five High-Promising Schools*, Harvard Education Press, Cambridge, MA, 2010, pp. 38–39.
10 See Gilligan, Carol, *In A Different Voice*, Harvard University Press, Cambridge, MA, 1982.

DOI: 10.1057/9781137386045

4

The Charter School Mess, A Messy Systems View

Abstract: *This chapter uses the Jungian framework to examine The Charter School Mess (TCSM) for several reasons. First, the strong differences present in every aspect of the debate between the proponents and critics of charters fall easily into different Jungian quadrants. Second, TCSM and TEM have not been analyzed as a bitter and prolonged struggle between sharply differing psychological worldviews. Third, resolving the innumerable surface policy issues is impossible unless one deals with the underlying psychological issues first. Fourth, the Jungian analyses of some of the many stakeholders in education reveal crucial differences between them that are not easily apparent from other studies. Fifth, these unresolved psychological differences are significant contributors to TEM. The Jungian framework helps us gain deeper insight into the issue of charter schools, why it is currently a mess, but need not be necessarily an irresolvable one.*

Mitroff, Ian I., Hill, Lindan B., and Alpaslan, Can M. *Rethinking the Education Mess: A Systems Approach to Education Reform*. New York: Palgrave Macmillan, 2013. DOI: 10.1057/9781137386045.

"West Oakland Middle School (WOMS) has been a 'failing' school for many years..."

"... WOMS, like many schools serving poor children throughout the United States, is overwhelmed by the social needs of the children. Approximately one in five children in the United States comes from families with incomes that fall below the poverty line. According to the Children's Defense Fund, the number of homeless children and youth in public schools in America has increased by more than 40 percent between 2006–07 and 2008–09. In most cases, even parents who can't find work or housing, or can't afford healthcare, can and do send their children to school. Consequently, the nation's public schools are shouldering the brunt of the economic crisis facing poor children, and they have done so largely without additional resources or even acknowledgement by state and federal officials."[1]

Pedro Neguera

General lessons from special schools

There is a very special set of exceptional and outstanding charter schools[2] that precisely because they are so special and outstanding—so different from and far above the "average"—contain general, if not universal, lessons regarding how to lower the persistent achievement gap between (1) mostly black and Hispanic and (2) mostly white students—and even more generally, between poor, urban disadvantaged and suburban, advantaged students.

As it always is, the "data" one uses, and especially how one interprets it, are key in the conclusions one reaches about charters, e.g., what works versus what doesn't. In other words, the data one picks consists of more than so-called raw test scores alone. Whether one is aware of it or not, as we indicated in our discussion of ISs in the previous chapter, the theory behind the selection and interpretation of the data are always integral parts of the data. From a system's perspective, data, theory, and their interpretation are inseparable, certainly from the standpoint of Systems Thinking.

The special set of charter schools to which we are referring include (1) five Boston charter schools that are the focus of in-depth case studies[3] and (2) the Harlem Children's Zone (HCZ)[4] that has not only been

DOI: 10.1057/9781137386045

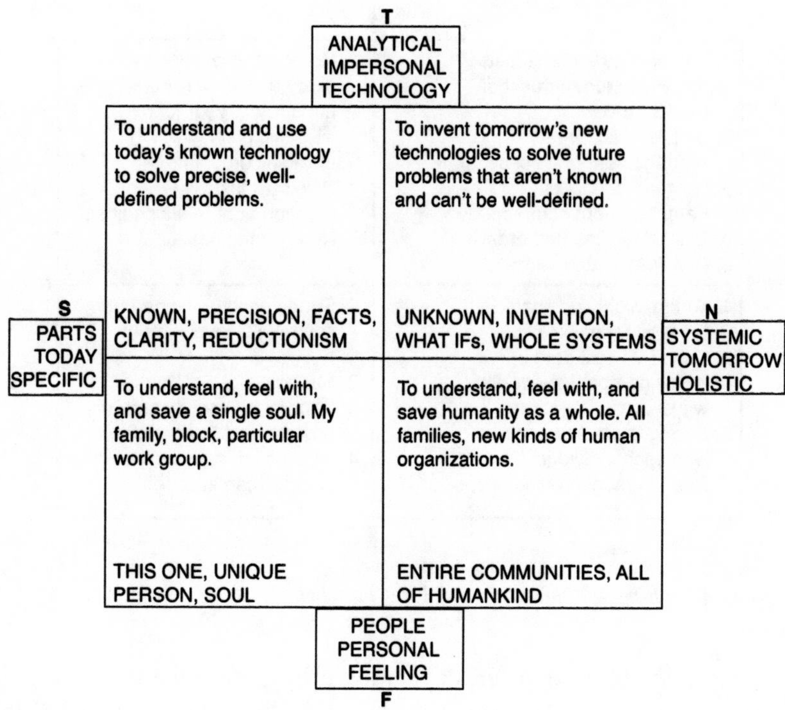

FIGURE 4.1 *The Jungian framework*

studied intensely, but singled out for widespread recognition and many awards.

Once again, Figure 4.1 shows the four primary Jungian quadrants that we discussed in the previous chapter. Figure 4.2 is not only a summary of what the special charter schools are doing in each of the Jungian quadrants, but it shows how all of the quadrants need to work together to reinforce one another. Strongest of all, the four quadrants are inseparable. *Success in any single quadrant cannot be achieved without a carefully orchestrated, concerted, and integrated policy for achieving success in all of the quadrants. In addition, success is any particular quadrant is dependent on achieving success with regard to all the items in that quadrant. This point alone differentiates this book from others with respect to charters, and to education in general.*

It is important to note that the authors themselves compiled the separate entries in each of the Jungian quadrants. Direct statements with

DOI: 10.1057/9781137386045

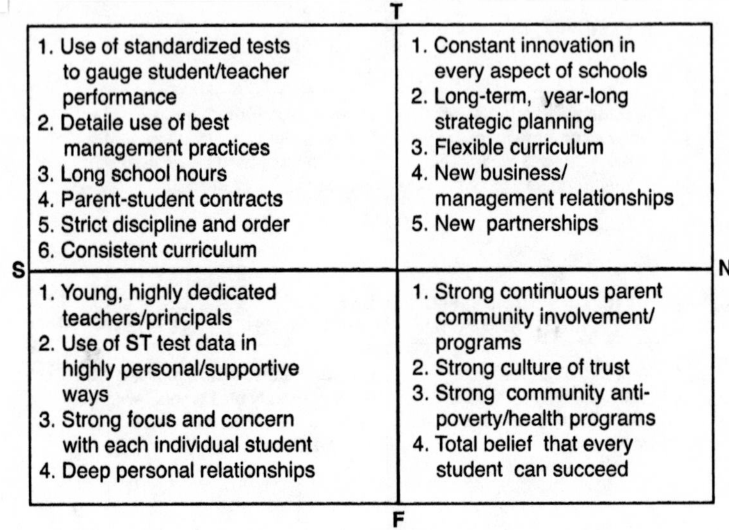

FIGURE 4.2 *Jungian types and special charter schools*

respect to the philosophy and operating principles of the special set of charters were assembled, interpreted, and finally coded as was done in the previous chapter, for instance, whether a particular statement best fit the Sensing-Thinking (ST) quadrant, etc.

From the standpoint of messes, it is extremely important to point out that each of the items in Figure 4.2 is simultaneously (1) a desired outcome or end, (2) an operating principle or means for achieving desired ends, and (3) a metric standard for evaluation. In addition, they are also (4) philosophical assumptions at the very same time. One of the major properties of messes is that every one of its constituent elements plays a number of roles simultaneously. As we said earlier, messes not only challenge but dissolve the facile distinction between causes, effects, means, ends, assumptions, etc. That's precisely why they are messes.

Recall from Chapter 2 that we hypothesized that *from a system's perspective, performance or success is multiplicative, not additive.* If it were additive, then the Jungian quadrants would be independent, which they are not. Expressed differently, the lowest possible score of 0 in any one quadrant is *not* offset by the highest possible score of 100 in any other. In short, 100 times 0 equals zero! Thus, *the total performance of* a system as a system *is a function of the product of the performances in each of the quadrants.*

DOI: 10.1057/9781137386045

Even stronger, the performance in any one quadrant is a strong "function" of the performances in all the other quadrants. Recall that we also acknowledged that at the present time, no one knows the exact nature of this "function" and that it varies depending on the particular mess.

No silver bullet

Figure 4.2 shows unequivocally that *there is no single "'magic' or 'silver bullet'" when it comes to reducing the achievement gap. Doing well in one or two quadrants is not an option for lowering the achievement gap. If anything, it is a prescription for perpetuating and reinforcing it. Significant efforts must be undertaken and sustained in each of the quadrants and they must work together seamlessly for any effort to succeed.* This is precisely why our notion of systems theory is so different from the typical ways it has generally been used and interpreted in education.

Figure 4.2 also shows that it is not a matter of doing well on one or two of the items in each quadrant. Each quadrant is itself a "system" of highly interdependent items.

This also reveals why in general the vast, overwhelming majority of reform efforts have at best concentrated on one or two quadrants, as well as one or two items per quadrant.[5] Without substantial philosophical and psychological training in systems and messes, it is generally too overwhelming and anxiety provoking to tackle efforts across the full mess. It also explains why most reform efforts have at best been of only mixed or partial success.

HCZ in particular shows that as the critics of charters have rightly contended that what goes on *outside* of schools is as important as what goes on *inside.*[6] This is why despite all the hype and rhetoric, *good teachers by themselves cannot lower the achievement gap. No single variable or intervention can.* The general conditions of poverty—poor housing, nutrition, health, living and working conditions—all contribute to the learning difficulties that poor children in particular—whether they are Black, White, urban or suburban—face in school.[7] For this very reason, HCZ has worked tirelessly to forge strong community support programs to help both parents and children overcome the debilitating effects of poverty. Thus, HCZ, along with others, offers classes on child development and good parenting, health clinics that are located directly in schools, job training workshops and fairs, housing and relocation programs, etc.[8]

DOI: 10.1057/9781137386045

In effect, Figure 4.2 is also a "management model for schools." It shows that "excellent management models and practices" are more different and far more complex than what has been wrongly portrayed by the critics of charters. In effect, they have used one-dimensional, highly flawed management models—in effect, some of the worst examples—to portray all that modern business and management practices have to offer to the better running of schools. Without their full knowledge or awareness, by incorporating key elements from each of the Jungian quadrants, the special set of charter schools have recreated the best management models of which we know at the present time for managing all of the critical tasks and functions that modern organizations have to perform successfully all the time. In addition, again without necessarily their full knowledge or awareness, the special set of charter schools has approached the problem (mess) of school design and management systemically. Whether they have been aware of it, they have adopted Systems Thinking in the form of the Jungian framework as their primary guide.

The point is that Figure 4.2 is not only a management model, but the best model of which we know at the present time. But, since all of the items are dynamic and not fixed for all time, we would be surprised if they did not evolve.

A highly interconnected system of very stringent conditions

The five Boston charters and HCZ show unequivocally the full set of actions, conditions, and interventions that must be put into place if one wants to lower the achievement gap. They not only embody the best set of conditions that are known at the present time, but make no mistake about it, they are extremely stringent. *They constitute a highly interconnected system such that if any one of them is missing, not implemented, or practiced to the fullest, then they will not work. In short, it's not piecemeal. It's all or nothing at all.*

No wonder most schools or reform efforts have not succeeded. It would have been a miracle if they had.

Figure 4.2 also shows the folly and futility of using any single, narrow set of measures, or metrics, to evaluate something so complex and messy as educational and school performance. If one collects standardized test data (i.e., ST) across schools of all kinds, especially since different

DOI: 10.1057/9781137386045

districts and states often use different standardized tests, then one would expect on average to find little if any differences between charters and public schools. One would in fact be surprised to find anything else. Conversely, if one only uses the first two Inquiry Systems—Expert Agreement and Analysis—to collect and analyze "data," then one would also be highly surprised to find anything to the contrary.

Interestingly enough, those studies of public schools that have done fairly well at closing the achievement gap show that they have also fallen short precisely because they have not pursued an integrated strategy based on the four Jungian quadrants[9]. If this is indeed generally the case, then is this why on average charters and public schools have shown no significant difference in closing the gap? Is the missing, crucial variable the "*degree* to which a school pursues an *integrated, systemic strategy*?" In other words, *the missing "variable" is not a single variable, but an integrated system of variables.*

Studies of effective public schools add needed support to the contention that a host of interlocking variables are responsible for raising achievement levels, and furthermore, that no single variable by itself can accomplish it. We quote:

> Edmonds showed that high achievement correlated very strongly with strong administration, high expectations for student achievement, an orderly atmosphere conducive to learning, an emphasis on basic skills learning, and frequent monitoring of student progress ... [10]

Notice that with the exception of "high expectations of student achievement," all of the other variables or factors are in the ST Jungian quadrant. However, the book from which the quote appears does show that Intuitive Feeling (NF) variables like school culture play an important role. Furthermore, while civic engagement and capacity are also recognized and singled out as important community variables, they are not treated expansively, i.e., systemically and systematically, as they are in HCZ, e.g., child, parenting, and job training classes; improved housing and medical care, etc.

Research also shows that if any quadrant is done in isolation, then by itself, it is not enough to lower the achievement gap. Thus, merely moving parents to better housing in better neighborhoods is not enough to make a substantial difference in all settings.[11] It is not that this doesn't make any difference at all, but in the case of charters versus public schools, on average, the difference, if any, is small. Furthermore, in many cases,

DOI: 10.1057/9781137386045

the variables are so hopelessly entangled that it is virtually impossible to separate them. We cannot stress enough that we are dealing with a complex, messy, ill-structured problem, not a well-structured one. We are dealing with messes, not simple systems that by definition can be partitioned into separate, simple effects and parts.

Furthermore, since we are always dealing with messes, we also cannot emphasize too much that the actions, conditions, etc. that need to be undertaken in each quadrant are not known or fixed forever. They are constantly changing as the environment changes. What works at one time in one setting will not necessarily work in others, or even in the same setting at other times.

In the above quote about Edmonds also makes the very important point that *all* of the factors *within* a particular quadrant need to work together, and not merely those *between* the quadrants. *The conditions for raising achievement and lowering the gap are daunting and severe, but they are not impossible. If there were NO schools and conditions under which lowering the achievement gap were possible, then the situation would truly be hopeless and impossible. That something is difficult—even exceedingly difficult—is not a proof that it is generally impossible. It is in fact a source, however tenuous, for hope.*

And of course, successful reforms take place within the even larger system of school districts, the states, and federal government, ideally all working together.[12]

For this reason, while we certainly believe that new technologies have an important role to play in the transformation of education, we do not believe that technology alone can lower the achievement gap. Even though we believe that the right kinds of technologies have the potential to play a significant role in transforming education[13], we still don't believe that technology alone can make up for all the variables that are needed, and especially their lack of integration.

As an aside, Mitroff had occasion to study a highly successful center that served the poor and homeless in a major American city. He found the same conditions that made for lowering the educational achievement gap. One of the major reasons why the particular center was successful is that it contained in house all of the services that were needed to help the poor and homeless, i.e., health, educational, job training, etc. As the director of the center put it, "Being poor is a full-time job!" The poor find it hard to find and keep jobs because so much of their day is spent in rushing around to different social service agencies that are supposed to help them. One does not help people whose lives are already too

DOI: 10.1057/9781137386045

fragmented by adding further to their overwhelming sense of frustration brought about by further fragmentation.

Lastly, we would be remiss if we did not say something about The Community Schooling Movement.[14] Make no mistake about it; The Community Schooling Movement is not only highly admirable, but a significant step in the evolution of public schools.

It incorporates many of the features of Figure 4.2. For instance, it forges deep connections between "outsiders" such as local and national businesses, organizations, and schools. For another, it brings health services directly into schools. What it doesn't and currently can't do under the mandate of public schools is to secure better housing and jobs for the members of impoverished communities as the HCZ does. Thus, a necessary and crucial part of a total systems approach to education is missing. If public schools can find a way to incorporate this crucial ingredient, then we believe that public schools can compete with the best of charters. Indeed, they must find ways to accomplish it.

The contribution of the expert disagreement IS

None of this means that one shouldn't use ST data and the Expert Agreement IS at all, for as limited as they are, they provide valuable information. But contrary to the Expert Consensus IS, data by themselves are not the final arbiter or absolute cornerstone of "truth." Data are "pointers" or "indicators," nothing more.

Data are an important component of all inquiry, but they are not the whole of it. Data are the inputs into a system whose output is a valid claim of an argument[15], e.g., "on average there is no difference between charters and public schools, or, x percentage of all children read at grade level, etc." First of all, data are never just data alone. All data are part of a broader category: Evidence. And, Evidence is the result of a prior inquiry. The values, methods, and theories that were used to specify and collect the data are fundamental parts of it. Once again, contrary to the Expert Consensus IS, data are not theory and value free. As result, data never "speak for and by themselves alone."

All of the Jungian quadrants collect and use a different kind of data. No one quadrant has a monopoly on data or truth. Thus, SF and NF data, which look at very special cases in highly personalized ways, are as valid as ST and NT, which look at impersonal, general, and abstract

DOI: 10.1057/9781137386045

data. As many have pointed out time and time again, it is not an "either/or" between *qual*itative SF/NF and *quant*itative ST/NT data but a "both/and." (Recall the discussion of SF Inquiry Systems.)

In the same way, each of the Jungian quadrants has a different sense and meaning of accountability. ST accountability is based on detailed, precise measures of performance according to carefully prescribed, generally accepted standards of evaluation via the Expert Agreement and Analysis ISs. NT accountability is based on what and how well a system is doing with regard to long-range, strategic planning (the Multi-Model IS). NF accountability is based on the culture of a system and how well it is contributing to the general well being of an organization and the surrounding community (Systems Thinking). SF accountability is based on how well students, teachers, and parents feel about themselves, their school, and community (Feminist Inquiry).

The unique AND the universal

This points to something even deeper, which is generally not appreciated. One of the biggest differences between T and F is that T looks at and emphasizes what is general and universal in human affairs. F on the other hand looks for and emphasizes what is unique. For this reason, it is not a fatal criticism of F—the criticism completely misses the mark—to say that F concentrates what is "unique" and "not replicable." F is not only fundamentally concerned with, but only recognizes that which is unique. Nonetheless, from a broader system's perspective, it is a valid criticism to say that neither T nor F by themselves covers all of the human condition. How could they?

From the perspective of the Disagreement IS of the previous chapter, "truth" is the result of the intense debate and interplay between T and F. As a result, they are more dependent on one another than they have dared imagine in their wildest dreams, or should we say, nightmares! In the same way, the proponents and critics of charters are more dependent on and need one another more than they have ever imagined.

Concluding remarks

The purpose of this chapter has been to present a systems overview, evaluation, and critique of charters that is very different from anything

DOI: 10.1057/9781137386045

of which we are aware and has appeared thus far. It is a jumping off point for the designs of future charter schools.

Finally, the analysis presented in this chapter suggests the design and execution of a new kind of research study. Metrics need to be developed for each quadrant of Figures 4.1 and 4.2 that help to measure what and how well schools are doing in each quadrant. Data then need to be collected on a nation-wide sample of different types of schools. Once this is done, multiple regressions and other more complex forms of statistical analyses can be performed to ascertain how much each Jungian quadrant contributes to the lowering of the achievement gap.

We cannot stress enough that in principle we are not opposed in the slightest to using ST data. Indeed, we'd like to see more ST data that are based on Systems Thinking. Indeed, we do not think highly of those who are opposed in principle to any and all application of ST.

Notes

1 Neguera, Pedro, "Stretching the School Safety Net," *The Nation*, January 2, 2012, p. 23.

2 See Merseth, *op. cit.*

3 *Ibid.*

4 Dobbie, Will, and Fryer, Roland, "Are High Quality Schools Enough to Close the Achievement Gap? Evidence from a Social Experiment in Harlem," National Bureau of Economic Research, Cambridge, MA, 2009; Tough, Paul, *Whatever It Takes, Geoffrey Canada's Quest to Change Harlem and America*, Mariner, New York, 2009.

5 Jennings, Jack, "Reflections on a Half-Century of School Reform: Why We Have Fallen Short and Where Do We Go From Here?," Center on Education Policy, Washington, DC, January, 2012.

6 Ravitch and Rothstein, *op. cit.*

7 Duncan, Greg J. and Murnane, Richard J. (eds), *Whither Opportunity: Rising Inequality, Schools, and Children's Life Chances*, Russell Sage Foundation, New York, 2011.

8 See Dobie, Will, "Are High Quality Schools Enough To Close The Achievement Gap? Evidence From A Social Experiment In Harlem," National Bureau of Economic Research, Working Paper 15473, Cambridge, MA, November 2009.

9 Cuban, Larry, and Usdan, Michael, *Powerful Reforms with Shallow Roots: Improving America's Urban Schools*, Teachers College Press, New York, 2003.

DOI: 10.1057/9781137386045

10 Datnow, Amanda, Lasky, Sue, Stringfield, Sam and Teddlie, Charles, *Integrating Educational Systems for Successful Reform in Diverse Contexts*, Cambridge University Press, New York, 2006, p. 13.
11 Duncan, Greg J. and Murnane, Richard J., *op. cit.*
12 *Ibid.*
13 Christen, Clayton, Horn, Michael, and Johnson, Curtis, *Disrupting Class, How Disruptive Innovation Will Change the Way the World Learns*, McGraw Hill, New York, 2008
14 "Building Community Schools," *The National Center for Community Schools*, http://nationalcenterforcommunityschools.childrensaidsociety.org/.
15 Toulmin, Stephen, *The Uses of Argument*, Cambridge University Press, Cambridge, UK, 1958.

DOI: 10.1057/9781137386045

5

The Charter Schools of the Future—Possible Designs

Abstract: *This chapter continues the use of the Jungian framework to analyze some of the other important stakeholders that bear on TEM. Specifically, it looks at the properties and roles of teachers and teachers' unions. The chapter shows that both teachers and unions are not equally developed in all of the Jungian quadrants. This lack of equal development significantly affects our ability as a society to lower the achievement gap between upper, middle class and poorer children.*

Mitroff, Ian I., Hill, Lindan B., and Alpaslan, Can M. *Rethinking the Education Mess: A Systems Approach to Education Reform*. New York: Palgrave Macmillan, 2013.
DOI: 10.1057/9781137386045.

> "The mentality of professionalism was reinforced by unions, which attempted to raise salaries and improve working conditions for teachers. While attempting to improve teaching as a profession, unions had the paradoxical effect of further imposing rigid work rules, occasional nasty and very public strikes, and other constraints on teachers' activities ..."[1]
>
> Sandra Waddock

Introduction

One of the most important uses and outcomes of the Jungian framework is that it allows us to give a more systematic treatment of the current strengths and shortcomings of charters. The Jungian framework not only gives a systematic understanding of the current limitations of charters, but it shows ways of responding to those limitations systematically *and* systemically. Most important of all, it gives a way of thinking about possible designs for future charters.

With regard to their current limitations, the critics of charters rightly point out that the groups and populations that opt out from public schools for charters in effect re-segregate themselves. Charter schools are even less integrated culturally, economically, racially, and socially than the public schools that have been left behind. This is not something that individual charters can necessarily undo by themselves. Mechanisms are needed whereby individual charters can be organized into larger integrated units that can respond more effectively to population and cultural imbalances. We therefore suggest one mechanism at the end of this chapter. In short, individual charters must experiment with and find ways of coalescing into larger units.

Once again, the fact that charters are even less integrated culturally, economically, racially, and socially than the public schools is less true now than it was at the beginning of the charter school movement. Nonetheless, the lack of cultural and racial diversity is still a problem.

For another, contrary to the critics, it is not clear that teaching to tests is wholly without merit. If it can be done in creative ways (in terms of NT, NF, Multi-Model, Disagreement, and Systems Thinking ISs), e.g., as jumping off points to more expansive discussions of broad topics, then there is nothing wrong per se in using tests (ST, Expert Agreement ISs,

DOI: 10.1057/9781137386045

etc.) as teaching devices. Furthermore, if tests are used in one-on-one ways (SF and Feminist Inquiry) to support and help individual students improve, then this strengthens the use of tests even more. Furthermore, because all tests can be "gamed," this does not mean that one shouldn't strive to design better tests that are harder to game. It surely doesn't mean that one should abandon all tests whatsoever.

Other critical stakeholders

One of the most important employments of the Jungian framework is its use to pinpoint critical differences between critical stakeholders, e.g., how they view, and contribute to The Education Mess. Indeed, the mismatch between different stakeholders is one of the most critical and important reasons for the existence of The Education Mess. While stakeholder differences have certainly been analyzed many times before, to our knowledge, they have not been commonly analyzed in terms of the Jungian framework.

In this chapter, we examine only two of the many critical stakeholders that impact education: new teachers and unions. Obviously, we are not contending that these two and these two alone either exhaust or are sufficient to analyze all of the stakeholders that impact and are impacted by TEM. Instead, these two show how in principle any set of stakeholders can be analyzed systematically and systemically.

Figures 5.1 through 5.4 show the forces acting on new teachers[2] and unions[3]. To put it mildly, they show that the separate Jungian quadrants have historically not always worked together, let alone well. They certainly do not always support one another even now. Furthermore, they are generally imbalanced, i.e., not equally developed in all quadrants. That is, the quadrants have not meshed within particular stakeholders, let alone between them. They certainly do not fit well with the stringent conditions that we laid out in Figure 4.2 and that we believe are vital to lower the achievement gap.

Let us examine some of the items in Figures 5.1 through 5.4 in greater detail.

In *Finders and Keepers: Helping New Teachers Survive and Thrive in Our Schools*,[4] Susan Moore Johnson makes the point repeatedly that as a general rule, teachers, new and old, are conservative, not necessarily in the political, but in the educational sense. (See Figure 5.1.) According to

DOI: 10.1057/9781137386045

		T		
	1. Teachers conservative, resist reforms 2. Unpredictable charter organization structures 3. Low pay 4. Low prestige 5. High need for structure 6. Want guidance & freedom 7. High turnover		1. Ill-prepared for new recruits 2. Teachers isolated, not interdependent 3. Old silo org forms treat teachers as interchangeable 4. Low pay 5. No energy, structures, time for strategic planning, thinking 6. Many other, new career options	
S				N
	1. Teachers conservative, resist reforms 2. Resist distinctions between peers 3. Personal commitment to social justice 4. Teaching life & death for kids 5. Rudeness, safety; school order 6. Collegiality 7. Meaningful career 8. Demoralization 9. Principal friendly, personal		1. Committed to social justice 2. Teaching own reward 3. New teachers enthusiastic about merit pay 4. Great achievement gap disparities 5. Effective schools have good personal relationships 6. Principal involved, peer feedback 7. Truly progressive 8. Induction seminars, retreats, mentoring 9. Social support services	
		F		

FIGURE 5.1 *Jung and the new generation of teachers*

		T		
	1. Teachers conservative, resist reforms 2. Unpredictable charter structures 3. Low pay a serious barrier to getting new teachers 7. High need for structure, order; both lacking 8. Little guidance; want guidance & freedom		1. Most schools ill--prepared for new recruits 2. Traditional schools isolated, not interdependent 3. Low pay a serious barrier to getting new teachers 4. No common planning time 5. No energy, time for strategic planning, thinking	
S				N
	1. Teachers conservative, resist reforms 2. Resist distinctions between teachers 3. Personal commitment to social justice 4. Resist distinctions between peers 5. Teaching life & death for kids 6. Constant struggle to find ways to help kids		1. Committed to social justice 2. Very few teach because of money 3. Teaching own reward 4. New teachers enthusiastic about merit pay 5. Great achievement gap disparities 6. Effective schools have good personal relationships 7. Low, poor feedback 8. Really progressive? 9. Induction seminars, retreats, mentoring	
		F		

FIGURE 5.2 *Jung and the old generation of teachers*

DOI: 10.1057/9781137386045

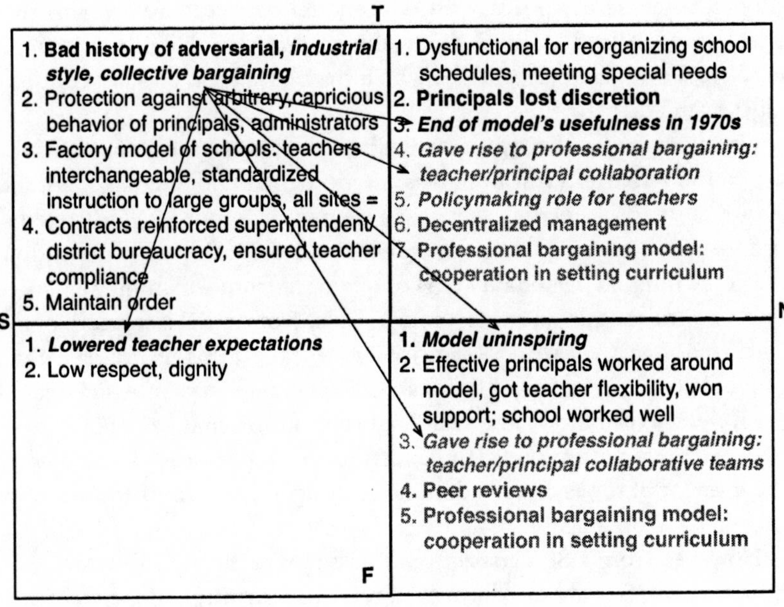

FIGURE 5.3 *Jung and unions—the old industrial model*

T	
1. Union benefits: more pay, lower class sizes, bargaining power, fringe benefits, protection against loss of employment 2. Slight positive effect on student achievement 3. Rely more on traditional instruction: focus on average student 4. Clear separation of roles 5. Increased cost of instruction	1. Neglect of special students 2. Long-term planning by administrators
S	N
1. Security, predictability 2. Demoralized?	1. Traditional school roles 2. Traditiional cultures
F	

FIGURE 5.4 *Jung and unions—a modern-day model*

DOI: 10.1057/9781137386045

Moore, teachers generally resist educational changes and reforms that upset standardized ways of doing things that they have been taught, practiced, evaluated on, and for which they have been and are continually rewarded.

Notice carefully that we have deliberately placed the item "teachers are conservative and resist reforms" in more than one Jungian quadrant. As we pointed out earlier, *any* single element or group of words, items, propositions, statements, or utterances varies considerably depending upon the Jungian personality type or quadrant from which they emanate or from which they are interpreted. Thus, from a ST perspective, the "fact" that "teachers are conservative and resist reforms" is precisely that, a "fact," i.e., the number of times and the concrete ways in which teachers have acted to block and resist reforms and/or changes in curricula, new teaching methods, etc. In fact, from an ST perspective, the greater the number of times one has done something, the greater the probability that one will do it again.

However, from a SF perspective, the issue is completely different. It is how teachers feel about educational reforms and why independently of all the ST data in the world with regard to how good or effective a set of reforms are that teachers still resist them in terms of their deeply held personal, individual feelings.

Many teachers, especially STs, are also upset by the unpredictable, new, and dynamically changing organizational structures of charters. This item can easily be placed in the SF quadrant as well. In fact, any and all of the items can be placed in any of the Jungian quadrants.

In contrast, NTs and NFs find the dynamically changing nature of charters especially appealing. For this very reason, we would not be surprised to find that a high proportion of NTs and NFs are especially attracted to charters. To our knowledge, this leads to another research study that has not been carried out, i.e., which Jungian types prefer which kind of schools. It also leads to the question, "How can charters be designed and redesigned so that they will appeal equally to all of the Jungian types?"

We shall not comment on the other items in Figure 5.1 since they are mainly self-explanatory. The only one we would emphasize is that from a SF and NF perspective, a critical factor is whether a principal takes a direct personal interest in his or her teachers. This means not only showing up to observe classroom performance and give feedback, but "continually being there for teachers," i.e., offering personal

DOI: 10.1057/9781137386045

support (SF, NF) for a job that by any measure is difficult to perform well.

In the edited volume by Tom Loveless, *Conflicting Missions, Teachers Unions and Educational Reform*,[5] the contributors examine the history of teachers' unions in the U.S. (See Figures 5.3 and 5.4.) A central point is that originally teachers' unions grew out of and were burdened by an adversarial, industrial, collective bargaining model. In this model, the interests of the "rank and file," i.e., ordinary workers, were unalterably opposed to and at perpetual loggerheads with the interests of management, i.e., those in higher authority. Prolonged conflict was thus the perpetual order of every day.

The black arrows in Figure 5.3 indicate that the adversarial, industrial, collective bargaining model led steadily to its own demise. Teachers were not the same as "interchangeable factory workers." They were professionals. To the chagrin of the U.S. automobile industry, it took much longer to recognize that factory workers were not interchangeable as well. Thus, slowly but surely, the adversarial, industrial, collective bargaining model gave way and spawned the professional bargaining model where teachers were treated as nearly as possible as equal partners in the "education process," not as "workers in an educational factory."

Idealized Designs

The concept of Idealized Designs is one of Ackoff's most creative and useful ideas. Idealized Designs challenge us to imagine new designs that embody all of the currently known, desired properties that we would like a system to have. Idealized Designs are not utopian for they must be capable of being implemented. That is, every Idealized Design must have an accompanying implementation plan.

Since virtually anyone is free to start a charter, then there are no valid reasons why entirely new and different sets of stakeholders, especially working together in various combinations, should not start them: (1) unions which in part have already started charters; (2) members of congress; (3) physicians; (4) social workers; (5) psychotherapists; (6) members of diverse professions, etc. In other words, one important way to overcome the differences between stakeholders is to have them jointly design new schools. This certainly does not mean that we somehow magically expect years of bitter history to be overcome overnight. But

DOI: 10.1057/9781137386045

it does mean that we have every right—hope—to expect that a truly innovative organization like HCZ will take a leading role and find ways to bring disparate stakeholders together to forge new designs.

As we have pointed out repeatedly, the ante is continually being raised. Messes do not stand still. They are not static. We believe that organizations like HCZ in particular now have a special obligation to reach out and work with even larger communities of stakeholders. Indeed, why shouldn't HCZ start an entirely new school of teacher education?

We take Figure 4.2 as our standard model for idealized design. In other words, the Jungian framework is not merely descriptive. It can also be used prescriptively as a model for idealized design. For this reason, we show it below once again. Once again to our knowledge, the Jungian framework has not been used this way previously.

As the basis for a prescriptive, idealized design, Figure 4.2 serves as a challenge. It certainly is not an accepted fact or way of doing things for many of the stakeholders that impact education. This doesn't mean that Figure 4.2 is locked into place for all time for as we have emphasized The Education Mess is constantly changing. Thus, the purpose of Figure 4.2 is to serve as a starting point for idealized design, i.e., for each stakeholder to ask itself, "How would we have to change so that the ideal that is reflected in Figure 4.2 would become a reality?"

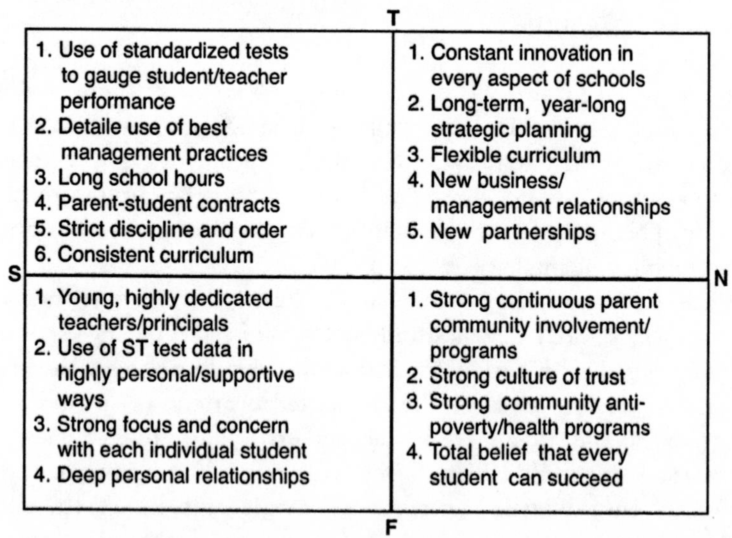

FIGURE 4.2 *Jungian types and special charter schools*

DOI: 10.1057/9781137386045

There is also no question that charters such as HCZ have to experiment with "ramping up." If they are not to remain separate, limited in scope, and isolated, they must find ways to cooperate and join together so that they can become equivalent to larger entities such as school districts.

HCZ is well positioned to work toward the goal of having at least one HCZ-like charter in each state. HCZ is well positioned to do this since it already offers workshops to various interested stakeholders regarding how to recreate the HCZ mission and design. We would be naïve if we thought that this alone would transform American education. For this reason, we turn to larger experiments that are unfolding at the state level.

But before we do, a final word is in order. In Mitroff's home office, he has a photo static copy of an 1896 letter from William James, arguably America's greatest philosopher, to the Provost of the University of Pennsylvania recommending Edgar Singer for a position in philosophy in that institution. In his letter, James says that Singer is the "best all-around student he has had in thirty years of offering philosophy at Harvard."

Singer went on to teach C. West Churchman with whom Mitroff studied the Philosophy of Social Systems Science at UC Berkeley. Thus, intellectually speaking, James is Mitroff's "great grandfather."

One of Singer's many contributions was his perceptive recognition of the difference between "goals or objectives" and "ideals." Goals and objectives are to be judged by how successful we are in obtaining them in a specified period of time. In contrast, ideals are things we perpetually strive toward, but never fully obtain, e.g., "lasting peace on Earth, the end of war, schools that are truly safe and secure, etc." The fact that we never fully obtain our ideals is not reason for despair or for their abandonment, for they serve as guiding aims of humankind.

The HCZ model, especially as we have analyzed it in terms of the Jungian framework, is an ideal that we perpetually strive to realize. For this reason we cannot specify if and when it will overtake and transform all of K-12 education. In our minds, it is an ideal eminently worth pursuing.

Notes

1 Waddock, Sandra, *Not By Schools Alone, Sharing Responsibility for America's Education Reform*, Praeger, Westport, Conn., 1995, p 15.

DOI: 10.1057/9781137386045

2 Johnson, Susan Moore, *Finders and Keepers, Helping New Teachers Survive and Thrive in Our Schools*, Jossey-Bass, San Francisco, 2004.
3 Loveless, Tom (ed.), *Conflicting Missions, Teachers Unions and Educational Reform*, Brookings, Washington, DC, 2000.
4 Johnson, Susan Moore, *op. cit.*
5 Loveless, Tom, *op. cit.*

DOI: 10.1057/9781137386045

6
Hiding in Plain Sight: Education Reform in Indiana

Abstract: *This chapter discusses the particular systems we know best. It thus discusses plans to revitalize the Indianapolis public schools and throughout the entire state of Indiana by means of pervasive education reform. In particular, we discuss the lessons that have been learned, and still need to be learned. Reviewing the lessons gives us another chance to witness the role that the Jungian framework has to play in evaluating complex social experiments.*

Mitroff, Ian I., Hill, Lindan B., and Alpaslan, Can M. *Rethinking the Education Mess: A Systems Approach to Education Reform*. New York: Palgrave Macmillan, 2013.
DOI: 10.1057/9781137386045.

Education accountability and reform: 1973–2013

Early 1970s in Indiana K-12 public education was a time of tumult and confrontation. Collective bargaining for public employees, and specifically for teachers, was working its way toward approval and was indeed approved in 1973. The movement was characterized by bitter conflict between school districts, administrators, and their managerial associations on one-side and teacher representative groups such as Indiana State Teachers Association and the American Federation of Teachers on the other side and the crucible for this struggle was the Indiana legislature. For decades prior to the seventies, public education governance in Indiana was paternalistic, local, interested in maintaining unilateral decision-making and determined to keep school taxes as low as possible. However, through the pressure of continued lobbying, job actions and strikes, the union movement won the day, but at a cost.

The quid pro quo underwriting mandatory collective bargaining was to shift significant control of public education to the state legislature. Immediately, school tax freezes were implemented to prevent "breaking the bank" by teacher unions and an aggressive program was initiated that still reverberates 40 years later: school accountability.

The State of Indiana's Curriculum C Rules

In 1975, the Indiana Department of Education (IDOE) introduced Curriculum, or "C" Rules which required school districts to forward copies of curricular objectives in key subject areas to the IDOE, as well as student performance with regard to the objectives, for state review. Educators from school districts across the state were outraged at the perceived invasion of their sovereignty. As a result, they packed IDOE informational hearings to register their anger. Hostility was palpable, language was confrontational, and IDOE presenters beat a hasty retreat. The backlash was so pervasive and lasting that IDOE capitulated and rescinded the requirements to report student performance. School districts only had to file copies of their curriculum objectives with IDOE.

What was in plain sight to government, business, and civic leaders remained generally unrecognized by the education community. It would continue to remain unrecognized for nearly 10 more years as education reform efforts remained dormant. Then, in 1983, another bombshell

DOI: 10.1057/9781137386045

rocked the country when Terrel Bell, U.S. Secretary of Education, released a study entitled "A Nation at Risk." The implications were obvious and onerous. Indiana Governor Robert D. Orr and Lt. Governor John Mutz took immediate action. They recruited Dr. H. Dean Evans, a former superintendent of schools, as State Superintendent of Education. At the time, Evans was president of the endowment fund of the pharmaceutical giant, Eli Lilly and Company.

The leadership of Lilly Endowment saw what others had not, that the quality of public education in Indiana was seriously deteriorating. They sought a leader who had the experience and charisma to put the storied and generous endowment squarely in the middle of education reform in Indiana. State government also needed an educational leader who could put the state's public schooling system on track. Evans more than accepted the challenge. Indiana's "A+" program was established when Evans' dominant piece, the Indiana Statewide Testing of Educational Progress (ISTEP), was put in place. It continues to this day, albeit with numerous modifications.

Evans worked tirelessly during his terms to make clear that Indiana's educational system was in jeopardy, to increase accountability, and improve student learning. Evans' successor, Suellen Reed, a longtime teacher, principal, and school superintendent was elected to the state superintendency. She continued efforts to improve schools by implementing one of the most aggressive accountability measures in the United States at that time.

Reed's program, Performance Based Accreditation (PBA), was a remarkable initiative based on solid research that was integrated with the realities of the effects of socioeconomics on learning outcomes. To avoid comparing poorer school districts with wealthier districts, PBA accounted for socioeconomic differences in school districts by clustering school into "leagues" based on the cognitive abilities of students and family income. In every school, student performance was measured against its peers.

PBA was not only an astonishing innovation, but it was equitable in its efforts to establish both high standards and high support for schools. Unfortunately, the educational community wanted none of it. As it had done before with C rules, it created such a backlash that by the late 1990s, Indiana's education landscape returned to a state of inertia that would remain for another decade. To be sure, there was occasional, lukewarm acknowledgement of the relentless deterioration in educational

DOI: 10.1057/9781137386045

outcomes. The education establishment in Indiana had an opportunity to be the first in the nation to see, identify, and *own* the education reform movement, but apparently, what was in plain sight was still hidden and stayed so for another generation of K-12 students.

In the midst of such denial, however, there were key people who saw the trends and they were alarmed. Bart Peterson, Mayor of the City of Indianapolis from 1999 to 2007, established the Office of the Mayor as a charter school authorizer and selected David Harris as first director of the charter school office. Despite entrenched opposition from the Indianapolis Public Schools and most public school leadership across the state, Mayor Peterson and Harris established the very first series of charter schools in the city of Indianapolis. The early days, where public and charter schools operated as if they were in "parallel universes," were fraught with non-starts, false starts, and complications. Nonetheless, the charter movement was not only begun, but it would not be turned back. As a result, it grew both in the numbers of schools and their quality.

Both Peterson and Harris would serve even more influential roles as the charter school movement gained acceptance and momentum. As the early work of Mayor Peterson and David Harris gained sustainability and credibility, they expanded their influence and extended their accomplishments in the education reform movement. Bart Peterson is currently Executive Vice President for Governmental Relations of the Eli Lilly Company and David Harris is CEO of The Mind Trust, an "education venture capital" firm of national renown that is central to a configuration of parallel schools that are emerging not only in Indianapolis, but the entire state of Indiana.

The current Mayor of Indianapolis, Greg Ballard, builds on the work of Mayor Peterson. He is broadening the scope and influence of the Mayor's Office of Education Innovation by means of a coordinated program to bring significantly more high quality charter schools to the city of Indianapolis. It is no coincidence that former Lt. Gov. John Mutz chairs the Mayor's Charter School Advisory Board. John is one of the key public servants who have the ability to see the severity of educational decline. He appreciates the implications of doing nothing, i.e., Absolving. Nonetheless, despite all the significant developments in Indiana, the growing influence of the federal No Child Left Behind Act, and its consequences for chronic school failure, the education establishment continued to miss what had been in plain sight for many years.

DOI: 10.1057/9781137386045

Fred Klipsch

There were of course many others who were very alarmed at what they saw. Few were more influential and motivated than Fred Klipsch, a longtime Indianapolis resident. Fred graduated from the Indianapolis Public High School College Preparatory Academy in 1959, went on to university, and became an extremely successful business entrepreneur. As his career in business grew, he watched in dismay as his alma mater went into free fall, was declared a chronically failing school, and finally taken over by the Indiana Department of Education.

Fred's business acumen and career success were and are matched only by his passion for shaping and improving the quality of education in Indiana. His involvement ranges from establishing charitable organizations for scholarships for low-income students who want to attend schools outside of their neighborhood to forming and supporting statewide education improvement groups. He envisions a "radically different type of school system" where schools operate according to a business-type model with "autonomous school systems" upholding high standards. This would mean significant reduction or elimination of central administrative offices and giving more freedom to individual schools, thus reducing bureaucracy and the role of government in education. This in turn would necessitate greater accountability to the local community and the involvement of parents.

To be clear, we are not opposed to the use of "business models" in education per se, but if and only if they are in accord with the best management models that treat students, teachers, and principals, and parents with dignity and respect, and most of all, support systemic efforts such as the HCZ. They must also recognize that although business models can certainly help improve schools, schooling is not a business.

In 2000, Klipsch became the chair of Choice Charitable Fund, a 501(c)(3) organization formed in 1992 that has raised more than $20 million in scholarships for students to go to private schools, parochial, charter, or other public schools away from a student's initial home district. The money gives students the option and the possibility of receiving an education that would be otherwise unavailable.

According to Klipsch, "free-market solutions" can make education more attractive for those teachers, parents, and students who have battled poor performance in their neighborhood schools to little advantage. (Again, we don't necessarily disagree but if and only if "free market

DOI: 10.1057/9781137386045

solutions" are done systemically as we have outlined in previous chapters. In this sense, we are anything but "unrestrained, free market advocates.) For this reason, he plans on implementing a broader program with educators and parents to teach them about Choice scholarships. Furthermore, Klipsch's plan includes expanding Choice scholarships from strictly income-eligible families to a universal program that allows any student, regardless of family income, to abandon their corresponding school or school district and use vouchers to attend higher quality schools. (We cannot stress enough that we are extremely skeptical about narrow, highly ideological so-called "free market solutions.".)

Such a policy has the potential to be a double-edged sword: more students may have the opportunity to attend schools with better standards and curricula or that fit the kind of learning students and their parents are seeking. On the other hand, increased student mobility could create instability in school systems where students constantly move from one school to another in search of the best school. Opening the program to students of all socioeconomic backgrounds could accelerate the departure of higher socioeconomic students with the family and financial means to travel farther to the best schools while lower socioeconomic students could be stranded in poor performing schools. A multi-tiered voucher with enhanced benefits for lower income families could serve to equalize educational opportunities for those students most in need.

Klipsch's deep-seated concern about public education is not confined only to children of lower income families. He is convinced that far too many children in Indiana have not received a good education for at least three decades and especially in the Indianapolis Public Schools. Among other staggering statistics, despite a growing budget, the quality of education in IPS has plummeted. 75% of African American students never graduate high school and 24% of 3rd and 4th graders are not able to read at grade level.

For Klipsch, public education costs too much and is mediocre at best. He argues that the fundamental cause of failing education in Indiana is the "legislative monopoly" that the Indiana State Teachers' Association (ISTA) holds on education. The teachers' union has consistently blocked legislation to improve education, in effect permitting only "the lowest quality service at the highest price." Moreover, the monopoly has protected the status quo of public education by maintaining control of it by serving as a legislative lobbying organism.

DOI: 10.1057/9781137386045

As the quality of education continued to deteriorate in Indiana, Klipsch recognized that the Choice Charitable Fund could help only about a thousand students per year versus what he saw as a pressing need for all of Indiana's one million-plus students. In 2007, Klipsch incorporated a 501(c)(4) organization to attack the lack of investment in education reform from the "education establishment". Klipsch's Hoosiers for Economic Growth (HEG) has lobbied for pervasive education reform through developing and launching strategies to change the model of education in the state.

Through HEG, Klipsch contends the current education structure is obsolete as evidenced by the United States' 25th place in math and 26th in science internationally. He argues that the system needs strategic reconsolidation to improve teaching and learning processes and thus change learning outcomes significantly. These improvements include changing teacher and administrator job descriptions to make them highly effective and efficient; measuring school progress by rigorous grading standards for both students and teachers; having a culture of high quality education through absolute clarity of most preferable learning objectives and the creation of high reliability school environments with lean budgets, high accountability and no room for mistakes. Under this model, local school district governance would be radically transformed; the Indiana Department of Education would delegate governance to for-profit and non-profit organizations. Hopefully, by embracing "free-market principles," incentivized school governance organizations would eliminate overhead costs, decrease inefficient building usage and limit the number of staff through innovative, technology-based blended teaching and learning. Teachers and school leaders would be contracted (or subcontracted), highly qualified, and compensated based on their performance.

HEG is arguably the single most influential organization in political and legislative engineering since collective bargaining for teachers was enacted into law in 1973. Many of the proposals developed by HEG since 2007 have been passed, most recently with the 2011 Indiana General Assembly.

Legislation now requires that measures of student achievement be based on the amount of learning growth occurring annually, teacher collective bargaining be limited to wage and benefits only, and teacher and principal performance be based in part on the amount of learning growth that occurs in each classroom and school. The Assembly also approved

DOI: 10.1057/9781137386045

a broad expansion of charter schools and created a publicly financed voucher program to give parents an option with respect to their choice of a child's school. According to the Indiana Department of Education (IDOE), "a voucher, or 'Choice Scholarship' is a state payment that qualifying families can use to offset tuition costs at participating schools. Schools qualify based on total household income and the amount of the scholarship corresponds with the public school corporation's financial profile in which the student lives." As mentioned before, the resulting "radically different school system" has more autonomy to design learning with much less bureaucracy, more autonomy, and significantly more accountability.

While hailed as a much-needed transformation by some, others in the field see it as a threat to educational accountability and even to the separation between church and state. The argument for transformation contends that privatization and individualization of schools would mean the end of accountability from traditional governance while others maintain that shifting tax monies from public to parochial schools fractures the line that separates church and state because those monies indirectly benefit religious institutions, especially the Catholic church because of its numerous schools.

There is legal precedent, however, for public taxes transferred to faith-based affiliated schools. As long ago as 1930, the beginnings of what would become the Child-Benefit Theory was first advanced "in support of a Louisiana statute providing for the appropriation of public funds for the purchase of school books for nonpublic school children in Borden v. Louisiana State Board of Education (1929)". The Louisiana State Supreme Court reasoned that the money would benefit the children and consequently the state. Thanks to this, and many others precedents, public money can be channeled to local private schools because the money follows the students from traditional public schools to those the parents choose. In Indiana, the Choice Scholarship provision of the 2011 legislation has been challenged in court by a coalition of litigants, including Indiana State Teachers Association and at least one national teacher union. Presently the program has been ruled constitutional:

> Arlington, Va.- Indiana's Choice Scholarship Program is perfectly constitutional. In a nutshell that was the ruling issued by Marion County Superior Court Judge Michael Keele in *Meredith v. Daniels*. The trial court rejected every legal claim brought by the plaintiffs—who are supported by both state and national teachers' unions—against the program, and it ruled in favor of

DOI: 10.1057/9781137386045

> both the state and two parents who intervened in a lawsuit in defense of the program. The Institute represents parents, Heather Coffy and Monica Poindexter, for Justice.
>
> Last August [2012], the trial court rejected the plaintiffs' request for a preliminary injunction against the Choice Scholarship Program until the court reached a final decision on the constitutionality of the program. Today's [June 2013] ruling is a final decision that marks the end of litigation about the program at the trial court level. It also means that the almost 4,000 students who have received Choice Scholarships can continue—without disruption—to be able to attend the private schools their parents have selected.[1]

Arguments by the plaintiffs that the Choice Scholarship Program improperly benefits private religious schools was rejected by the Court. The court maintained that the Choice Scholarship Program provides financial scholarships to the recipients directly for education at any type of school setting, whether public, secular private, or religious private.

A shift

In 2008, Governor Mitch Daniels, newly elected State Superintendent of Schools, Dr. Tony Bennett, and others, including Fred Klipsch, took decisive action. As a result, educational reform in Indiana vaulted to the forefront nationally. Unwilling to tolerate what they perceived to be continuing educational myopia and abiding comfort in the status quo, they orchestrated the most dramatic and comprehensive education reform legislation in the United States. A recent study conducted by the education research-oriented Thomas B. Fordham Institute identified national frontrunners in reform-minded states. By their standards of measure, they identified Indiana as the "reformiest" state in the nation:

> In what has been a monumental year in education reform for many states, Indiana has seen the most impactful and far reaching reforms passed and enacted. No one has been more successful in providing a more comprehensive reform plan for a system that is failing America's children.... New laws regarding teacher quality require local districts to adopt compensation models based not on teacher seniority and credentials alone—but on teacher effectiveness, leadership roles, and the academic needs of the students, as well. Schools must develop and implement fair, locally developed multifaceted annual evaluations for teachers and principals that consider student

DOI: 10.1057/9781137386045

> achievement and growth. Also included are personalized, meaningful professional-development plans and goals for teachers and principals that are informed by these evaluations. Indiana is even developing an evaluation tool that meets these criteria and will be available to school that want to adopt it.[2]

The new model for K-12 public education is centered on the key understanding that a centralized, top-down approach to education is ill advised and unlikely to produce success. Instead, the vision for K-12 education recognizes that a decentralized, school-based approach is key to educational excellence. In the parlance of Dr. Bennett's vision for the Indiana Department of Education, "Competition, Freedom, Accountability", Indiana's program establishes internationally competitive standards, gives schools maximum flexibility in operations, creates multiple school options ("parallel universes"), and holds all schools accountable. This requires strong, capable K-12 school leaders to ensure that students are being educated to their greatest potential. It also requires that content-strong teachers are receiving the support they need and that financial investments are made available to support outstanding classroom processes to the maximum extent possible. Additional components in Indiana's blueprint for education reform are the need for transformative school leadership and leadership development, a comprehensible and equitable A-F school grading system for identifying high-performing, marginal and failing schools, the Indiana Growth Model, which measures student learning in terms of growth (and in part, also estimates teacher effectiveness, and subsequent compensation, as it is correlated with the growth in student learning). It also features the RISE Teacher Evaluation and Development Program that all school leaders must implement through annual teacher evaluations.

International competition is another reality. School leaders, principals, and teachers, must be aware of the Organization for Economic Co-operation and Development's (OECD) Programme for International Student Assessment (PISA), its correlation with the United States' use of the National Assessment of Educational Progress (NAEP), and develop durable strategies for improving the international competitiveness of Indiana schools.

There are additional provisions in the education reform movement. The Indiana Department of Education is pushing aggressively for the expansion of charter schools throughout Indiana. To help further this goal, it has created the Indiana Charter School Board to serve as the

DOI: 10.1057/9781137386045

IDOE charter school authorizer, a move that signals strong commitment to developing a parallel set of schools.

By 2012, the educational establishment was beginning to realize that, if they could not or would not acknowledge what was hiding in plain sight, significant numbers of citizens, including the majority of elected representatives to the Indiana House of Representatives and Indiana Senate, had not only seen what was in plain sight, but had passed colossal and radical education reform measures that left the education establishment in near total disbelief and complete disarray.

The ongoing backdrop to the drama is the daunting and depressing fact that Indiana's largest school system, Indianapolis Public Schools, still contains some of the state's worst-performing schools. Reading, math, and graduation rates are shockingly low, if not abysmal. In well-documented cases, graduation rates in some Indianapolis high schools inched upward only because nearly 50% of graduates were granted waivers from passing the graduation qualifying examinations. In other words, the students did not actually qualify for graduation on the basis of academic achievement, but were simply waivered in order to increase the graduation rate. By early 2012, five schools in Indiana, four of which are in the Indianapolis Public Schools, were "taken over" by IDOE and assigned to private and/or proprietary Takeover School Operators (TSO). Since TSOs don't believe that the public schools can deliver, they have turned to charter schools as the primary way out. In effect, they have given up on the public schools. If only for the reason that children need a common, shared experience to become responsible citizens, we do not necessarily agree with this. We continue to remain strong proponents of public education. But make no mistake about it; all schools need to change.

The Jungian framework

The preceding discussion would make it appear that we are staunch advocates of everything that is happening in Indianapolis and the State of Indiana. While we are certainly staunch proponents for radical change, we are not uncritical of many of the major proposals. Our use of the Jungian Framework to analyze and explore plans that have been proposed at the city level of Indianapolis and the state of Indiana helps makes this clear.

DOI: 10.1057/9781137386045

While the city plans, proposed and led chiefly by the Mayor's Office and The Mind Trust, and the state, led chiefly by the State Superintendent of Education, embody many of the systemic features that we argued in Chapter 4 that were absolutely necessary for the success of charters in lowering the achievement gap (see Figure 4.2), as a general rule, they are seriously deficient when it comes to community involvement (NF) and in integrating all of the Jungian quadrants. Most of all, they are seriously deficient when it comes to incorporating mechanisms to combat and improve the conditions of poverty (poor health, medical care, diets, housing, parental jobs, and general living conditions, etc.) that especially inflict the majority of children in Indianapolis public schools. For this reason, despite the best intentions of the plans and their proponents, we are extremely skeptical as to whether they will succeed in turning Indianapolis and Indiana schools around. In short, they are largely premised on administrative, economic, structural, and teacher changes and reforms. While these are crucial, they are not systemic enough.

Figures 6.1 through 6.4 represent the essence of The Mind Trust's plans in terms of the Jungian framework.

	T		
S	1. Prove model 2. Support/undertake research 3. $188,000,000 controlled by Indy Public Schools; if reallocated to school leaders, then accelerate learning and operation of schools 4. Youngest children guaranteed early start 5. Only 45% pass Eng language arts, math; only 50% graduate 6. 6 out of 7 most chronically failing are Indy Public Schools	1. Empower ed entrepreneurs 2. Overcome barriers to success 3. Expand impact 4. Most innovative reform in Indiana; drive systemic reform 5. Transform Indy schools management relationships 6. Creative educators call shots 7. Time to think big 8. Instead of fixing curren system, unleash creativity to create opportunity schools where students learn	N
	1. Every child has excellent teachers 2. Empower parents 3. Meet family needs, stay open longer 4. Most principals have littile say who is on their team 5. Give teachers more say in what and how it gets taught 6. System not people is broken	1. Empower ed entrepreneurs 2. Overcome barriers to success 3. Every child has excellent teachers 4. Low income students make inspiring progress 5. Improve public ed for underserved students 6. Develop relationships between key stakeholders, new partnerships	
	F		

FIGURE 6.1 *The Mind Trust, part 1*

DOI: 10.1057/9781137386045

T

S			N
	1. Plan based on extensive analysis, research 2. Eliminate top-down district regulations that control curriculum, staffing, budgets 3. If schools don't perform, close 4. Schools accountable for own results 4. Clear separation of roles 5. Pay teachers more for great results 6. No new taxes; can do with current funding	1. New schools flourish 2. Eliminate top-down district regulations that control curriculum, staffing, budgets	
	1. Indy Public School Board makes reform/change impossible 2. Use technology for personalized learning	1. Coordinate city services; support students and families 2. Build climate for innovation reform 3. Parents choose from variety of outstanding neighborhood schools 4. Indy national magnet for most talented teachers, principals entrepreneurs 5. Opportunity schools unique to transform Indy Public Schools, lives, city's future	

F

FIGURE 6.2 *The Mind Trust, part 2*

T

S			N
	1. Coherence, focus 2. Schools in better position decide spend $$ to educate their students 3. Audit all Indy Public Schools to uncover savings 4. Benchmark 5. Pay teachers more for great results 6. Special need students score even lower than average	1. Philosophy, values guide school decisions; foster distinctiveness, coherence, focus 2. Develop new arts, foreign language programs 3. Hire 8-10 transformation directors to turnaround 8-10 schools 4. Systems coordinator 5. Network of outstanding charters	
	1. Parent buy-in essential 2. High quality preschools boost learning for life	1. Indy has legacy of civic engagement, community pride 2. New generation of top teaching talent 3. Unique mission aligned with schools 4. Philosophy, values guide school decisions; foster distinctiveness, coherence, focus 5. Principals create unique programs 6. Parent buy-in essential 7. Develop new arts, foreign language programs	

F

FIGURE 6.3 *The Mind Trust, part 3*

DOI: 10.1057/9781137386045

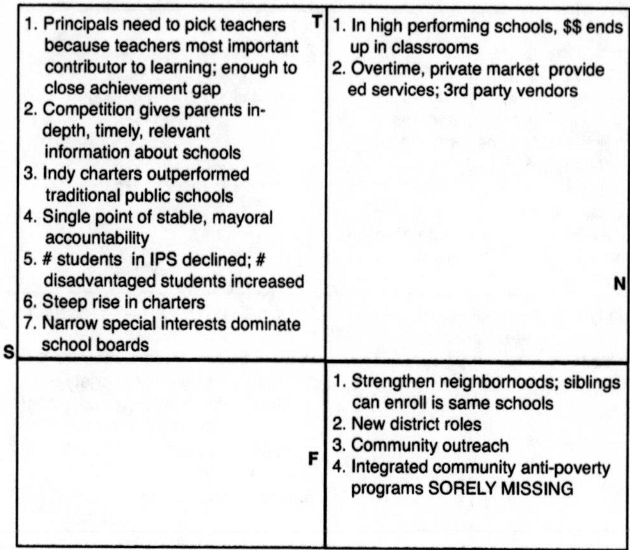

FIGURE 6.4 *The Mind Trust, part 4*

Figure 6.5 and 6.6 represent the state's plans.

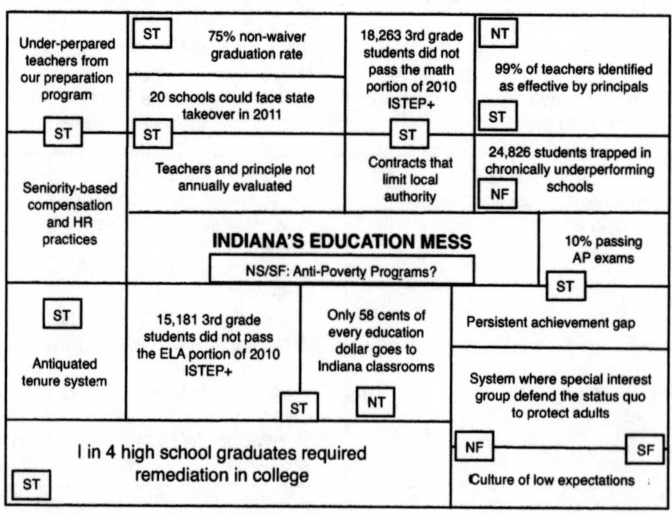

FIGURE 6.5 *Indiana's Education Mess, part 1*

DOI: 10.1057/9781137386045

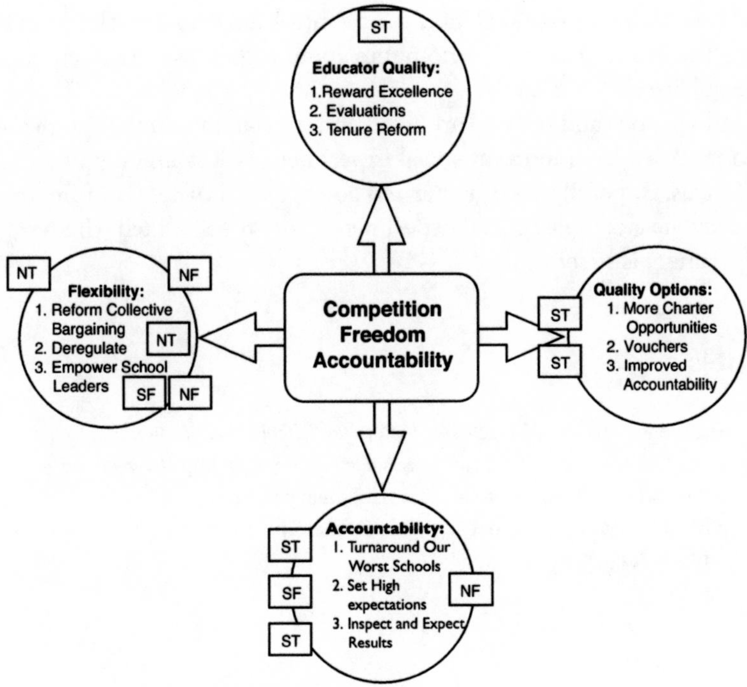

FIGURE 6.6 *Indiana's Education Mess, part 2*

Notice carefully that the central label in Figure 6.5 is "Indiana's Education Mess." Mitroff was honored to have State Superintendent of Education Tony Bennett in the audience for his kick-off lecture for a new program at Marian University for training new kinds of public school principals. Mitroff was there because Systems Thinking is a vital part of the new program.

Unfortunately, the research for this book was not performed at that time. If it had been, hopefully it could have made a significant difference in the thinking of the State Superintendent of Education and his staff.

Concluding remarks

Indianapolis and Indiana are important and highly instructive. In Indianapolis, there is a "war" going on between (1) an entrepreneurial, private, non-profit group, The Mind Trust, (2) the mayor, and (3) the state as to which party shall take over, control, and run the schools in the city. The Mind Trust makes no bones about it that it wants to eliminate the public schools entirely.

Indianapolis and Indiana are small and yet complex enough so that we can track a very important social experiment as it is literally unfolding before us. Hopefully, this chapter and book give a broader set of "metrics" to evaluate a complex social experiment. Only time will tell whether the experiment is successful.

Notes

1 Institute for Justice, Web release, *Victory for School Choice in Indiana, Court Rejects Challenge to Indiana Choice Scholarship Program*, http://www.ij.org/indiana-school-choice-release-1-13-2012, January 13, 2012.
2 "Education Gadfly" internet news notes from The Thomas B. Fordham Institute, August, 2011.

DOI: 10.1057/9781137386045

7

General Heuristics for Coping with The Education Mess

Abstract: *As we have emphasized repeatedly, messes are not well-structured problems that admit of simple solutions. As a result, there are no algorithms, methods, or fixed, ironclad rules that guarantee exact, permanent, and stable solutions. All one gets are heuristics that allow to cope as best as one can. This chapter discusses several such heuristics. If there were no ways of coping at all, then the situation would be entirely hopeless. The fact that there are heuristics argues that while the situation is daunting, it is far from hopeless.*

Mitroff, Ian I., Hill, Lindan B., and Alpaslan, Can M. *Rethinking the Education Mess: A Systems Approach to Education Reform*. New York: Palgrave Macmillan, 2013. DOI: 10.1057/9781137386045.

> ...a new open-admission public school called P-Tech, or Pathways in Early College High School...[launched] last September...[is] a partnership of the New York City department of education, the New York City College of Technology, the City University of New York and IBM, whose head of corporate social responsibility, Stanley Litow, used to be the city's deputy schools chancellor. The goal is to create a science- and tech heavy curriculum that prepares kids...for entry- and midlevel jobs at top tech-oriented companies. Each student gets an IBM mentor from Day One. There's a small but serious core curriculum focused on the basics: English, math, science and technology.[1]

Introduction

To see how easily and quickly education becomes a complex messy system (TEM), consider this. In terms of the Jungian framework alone, and for only three of the many major stakeholders, there are 81 different types of potential interactions between 1) principals, 2) teachers, and 3) unions. There are three possible choices for the ST quadrant, three for NT, three for NF, and three for SF. Thus altogether, there are $3 \times 3 \times 3 \times 3 = 3^4$ or 81 different combinations of possible potential interactions between the Jungian quadrants of principals, teachers, and unions.

If in addition we take the direction of interactions into account—e.g., stakeholder A may be influenced by B, but not necessarily B by A, i.e., the interactions can be either be unidirectional or bidirectional—then there are 81 times 3 or 243 possible kinds of interactions. (There are three because 1. A can be influenced by B, but not B by A; 2. B can be influenced by A, but not A by B; and 3. A and B can both be influenced by one another.) No wonder education easily and quickly becomes a mess! And, this doesn't even take into account the fact that there are countless other frameworks for describing and analyzing the different types of possible interactions and relationships between different stakeholders. (This conclusion is perfectly general for all messes since all of them contain an almost innumerable number of stakeholders.)

The general formula is this: if we have n different stakeholders, then there are $3 \times n^4$ possible types of interactions between their Jungian quadrants. To say this stretches the bounds of coping with messes is a

DOI: 10.1057/9781137386045

gross understatement! And yet, we have no choice but to cope as best we can. We do not have the luxury of absolving messes, i.e., doing nothing in the hope that "it all will just go away!"

TEM is here to stay no matter what we do. This doesn't mean that there are not good and bad ways of coping with it and for messes in general, for there are.

In this chapter, we present a variety of heuristics, i.e., broad "rules of thumb," for coping with messes. As stated previously, we only have heuristics for the basic reason that messes are ill structured. There are no fixed, permanent, or ironclad rules that guarantee exact and complete solutions of all messes for all time. The best we can do is to cope via various "rules of thumb." With this in mind, let us list the various heuristics with an eye of not only illustrating as many of them as we can, but showing how they apply to TEM.

In the next, and final, chapter, we explore many of the heuristics in-depth when examine the issue (mess) of school safety and security.

Key heuristics for coping with messes

Fundamental Preconditions

1 First of all, recognize that everything now needs to be treated as a mess in its own right or as part of a mess.
2 Recognize and accept that treating problems with a Machine Age mindset only makes Systems Age problems worse. One cannot even begin, let alone proceed, without these first two.

Increase the Diversity of Perspectives

3 View a mess from as many different perspectives as possible. For example, look at TEM not only from a financial or economic perspective, but also from a psychological, sociological, anthropological, historical, moral, political, technological, and spiritual perspective, among many. In each perspective, find at least one producer of the mess. (A "producer" is a necessary factor for the "production" of a mess, but by itself, it is "not sufficient" to cause a mess. In other words, a "producer" is one of many "co-producers." For example, planting an acorn is necessary to "produce" an oak tree—an end "product"—but other co-producers

DOI: 10.1057/9781137386045

such as air and water are also necessary. A "producer-product" relationship thus stands in sharp contrast to a "cause-effect" relationship where a prior "cause" is both necessary and sufficient for an end "effect.") Accordingly, human cognitive biases (psychological), the culture of schools and education in general (anthropological), the sheer and persistent politics of education (political), and so on, all play a major part. Don't worry if the producers do not fall neatly under any single scientific discipline or profession because they won't. Next look at the consequences of the mess. Ask, "What are, and what will be, the consequences of The Education Mess?" Don't focus on any single one of the consequences alone. Instead, try as much as possible to look at the economic, psychological, sociological, anthropological, moral, political, technological, spiritual, historical consequences, etc. in conjunction with one another, i.e., as systemically as one can.

4 Never ever trust a single formulation of a mess. Seek out and sweep in the analyses of those experts who cross-connect fields. Get different stakeholders from different professions to formulate a mess. For instance, long before 9/11, the artist Mark Lombardi developed intricate and elaborate ways of uncovering and tracing complex webs of international financial corruption. Part investigative reporter, postmodernist art historian, and graphic artist, Lombardi showed that by turning to public sources of information, he could demonstrate convincingly that the bin Laden and the Bush families were connected through a complex series of nefarious financial dealings. In short, Lombardi developed a new art form that showed pictorially how disparate and powerful global actors were interconnected. He showed the seamy side of the global economy. As a result of his work, Lombardi was one of the few artists to be accorded the dubious distinction of having his work examined by an FBI agent, in a museum no less, in order to gain clues into the terrorist financing of 9/11.

Examine and Challenge Taken-for-Granted Assumptions and Beliefs

5 In particular, using the fields of psychoanalyisis and clinical psychology, examine the deep and thereby often unconscious assumptions that are made about different stakeholders. It is not that stakeholders are "completely irrational." They are not "perfectly rational" in the conventional sense of rationality, i.e., in terms of

DOI: 10.1057/9781137386045

Analytical Rationalist ISs. This also increases the diversity of inquiry by forcing us to put ourselves in the shoes of different stakeholders. Because no stakeholder is ever perfectly rational or irrational, every stakeholder's perspective is at least partially rational.

For instance, Ackoff once talked about the illiteracy problem in a Black ghetto where programs created to solve the problem failed repeatedly. As a result, many concluded that the children were uneducable. But Ackoff did not think so; he assumed that the children were rational: they didn't want to accept "Whitey's" values (and thus invite abuse from their peers). Ackoff showed the children silent movies with subtitles, thereby providing the motivation to learn how to read.

For another, in *Why Some Politicians Are More Dangerous Than Others*[2] NYU psychiatrist James Gilligan makes an extremely powerful case in explaining how and why lethal violence, whether in the form of homicide or suicide, has historically increased significantly under Republican presidents and declined just as significantly under Democratic presidents. Indeed, lethal violence regularly reaches epidemic levels under Republican presidents. Even under Democrats, it is still way too high compared to other developed nations.

The link is as follows: While Republicans perpetually talk about getting tough on crime, they actually need high crime rates to get and stay in power. Because Republicans are strongly motivated in helping the rich maintain their wealth, pitting the lower middle class and working poor against the chronically unemployed and unemployable poor—who are seen as the primary parties largely responsible for crime—is a great way of diverting attention away from the fact that unemployment, income, and social inequality—all of which are major factors responsible for crime—actually increase substantially under Republican presidents. This is precisely why Gilligan sees some politicians, mainly Republicans, as more dangerous than others. Unfortunately, there are enough dangerous Democrats to go around as well.

More importantly, as a psychiatrist, Gilligan digs deeper for the underlying unconscious elements of human behavior. Republicans, and the Red State constituents they represent, are governed largely by a shame-based morality or ethic. Democrats, and their Blue State constituents, are governed largely by a guilt-based morality.

Shame is the deep, persistent feeling that, "I am bad." On the other hand, guilt is the feeling that, "We or I did something bad, but we are

DOI: 10.1057/9781137386045

not necessarily or inherently bad ourselves." Suffering shame, by say being fired or chronically unemployed, often leads to feelings that one is irredeemably bad to the depths of one's core. This in turn often leads to powerful feelings of wanting to strike back with intense acts of violence against others (homicide) or oneself (suicide). Whether "they" or one's self is actually responsible is besides the fact for in the wounded psyche everything and everyone is bad and therefore at fault. Red States intensify such feelings because they have a culture that inculcates and legitimates violence, and therefore, in subtle and not so subtle ways encourage its use.

In contrast, under guilt, one is motivated to help those who through no fault of their own have suffered, e.g., racial discrimination, unemployment, etc.

6 Monitor different stakeholder assumptions over time so that as the assumptions change, one can show the corresponding changes in how various messes are conceived and represented. If messes are the new reality, then assumptions are the building blocks of messes, and hence, of reality. As assumptions change, different perspectives on reality emerge. In this sense, reality is constantly being constructed and reconstructed over time. Also, a crisis occurs when all or nearly all of one's basic, taken-for-granted assumptions collapse. Thus, what assumptions are most vulnerable? Which ones are believed to be invulnerable? What are an individual's, an organization's, an institution's, or a society's crisis plans, if any, for what to do in the case where its major assumptions collapse? Thus, for instance, as teacher's unions changed their forms of collective bargaining, their underlying assumptions changed about a host of stakeholders that affect education.

Visit/Examine Extremes; Perturb the Ordinary/Conventional

7 Imagine/Design the Impossible. Ackoff's idealized redesign frees us from thinking that we are forever bound by today's constraints. Imagining and designing the impossible not only frees us from today's constraints, but it also forces us to question our deepest assumptions, e.g., "Why are today's constraints forever to be constraints?" Thus, Teach for America challenges traditional ways of preparing teachers. Sugata Mitra[3] questions a fundamental and conventional assumption we hold about education: students need

DOI: 10.1057/9781137386045

educators. In his "Hole in the Wall" experiments, Mitra shows that motivated by curiosity and peer interest, students may learn better in the absence of teachers.

8 Ask "Smart-Dumb" Questions. Never accept conventional, traditional constraints, or boundaries. Always have someone play the devil's advocate. Even more important, using the Expert Disagreement Inquiry System, construct a Dialectical opposite to the primary Inquiry System in use. For example, why can't students be "test-makers" as well as "test-takers?" If students participated in the construction of tests that measured their progress, would they thereby learn more effectively what they need to pass the tests? Would it help or hinder learning if teams of parents and students participated in constructing tests? Do students need educators (see #7 above)? Are we preparing students for the jobs of yesterday? Do educators know anything about the jobs of tomorrow, 10 years in the future?

9 Pay special attention to outliers. An outlier is an event or observation "that appears to deviate markedly from other members of the sample in which it occurs"[4], or "a person or thing situated away or detached from the main body or system" (New Oxford American Dictionary). Because what we observe is a function of our theories, outliers often inform us more than what we expect to observe. For instance, if we find too many outliers, then this may indicate that our perspectives are too narrow, for instance, by putting different things/people in too few or the same categories. Thus, one needs to sweep in more perspectives to make sense of outliers and messes. For instance, are Bill Gates and Steve Jobs outliers? What explains their successes? What is the role of education in their successes (and failures)? Malcolm Gladwell's *Outliers: The Story of Success*[5] provides an interesting perspective on the purposes of education: to standardize, to create outliers, or perhaps both.

10 Use "random interventions." These are deliberate strategies designed to understand the "noise" in any system. Noise is that which one cannot make sense of. Noise may include outliers and more. Another way to view it is: *A mess is misunderstood order, and order is a misunderstood mess. A mess is a kind of order that is not yet understood as such.*

DOI: 10.1057/9781137386045

In other words, messes are not totally devoid of order. They are a different kind of order. *Their "order" consists of the interactions between the "parts."*

Understanding or making sense of the "noise" in a system requires a great number and variety of different perspectives. What is noise according to one perspective may as well be order according to another. In fact, messes and order are opposite sides of the same coin, i.e., reality. The point is that "complete order" and "complete messiness" are the end-points of a continuum. Each concept makes use of the other in its definition.

When it is not possible to create random interventions, at least learn as much as possible from the misfortunes of others. For example, in 2005, Hurricane Katrina devastated Tulane University, forcing them to reconsider what is noise and what is essential to a university. Tulane went through a massive reorganization. They reconsidered their priorities, eliminated programs, simplified inefficient organizational structures, and even created new programs. It is now a smaller but a more focused university.

What can other schools learn from Tulane? How does the new Tulane compare to the old one in terms of performance criteria such as student placement, academic publications, research grants, etc.?

Investigate/Understand the Complexity of Interactions; Examine Improbable Interactions and Stakeholders

11 Ask at least two questions: 1. "What are some of the problems and messes that 'produce' a particular mess?" 2. "What are some of the problems and messes that a particular mess 'produces'?" In other words, go forward or backwards in time to connect problems and messes. The key point is that, in systems terms, problems are co-produced by other problems. From a systems point of view, it is not only safe, but vital to assume that all problems or messes are linked with other problems and that no problem can be solved or formulated in isolation from other problems or messes. For example, the educational problems of a country cannot be formulated, let alone be solved, in complete isolation from other problems. To repeat, the "co-producers" of a problem (which are themselves problems produced by other problems) are necessary, but by themselves, are not sufficient to result in a mess.

12 In messes, the interactions between parts, not the parts themselves, are the fundamental topics of investigation. Therefore, design specific scenarios that deliberately probe for difficult interactions. A seemingly "simple" example from medicine is instructive.

DOI: 10.1057/9781137386045

Janitors are one of the key stakeholders in keeping hospitals free from infection, not just nurses and doctors who treat patient infections directly with antibiotics, etc.[6] Since many "infectious bugs" adhere themselves to the metals in hospitals, e.g., beds, pans, etc., janitors are key in cleaning items and places that nurses and doctors don't "treat."

- **a.** Give special attention to the most improbable interactions, whether they are important, i.e., consequential, or not. These are the ones most likely to cause major crises. In fact, every major crisis has been shown to be the result of two or more assumptions, factors, interactions, etc. that were assumed to be unlikely, etc.
- **b.** Look at what seem to be least important interactions. These deserve special attention for these are the ones that come back to haunt us.
- **c.** Look at the most damaging interactions.
- **d.** Pay special attention to counter-intuitive, paradoxical, and unintended interactions and relationships. For example, in *Republic Lost*,[7] Harvard Law School Professor Lawrence Lessig notes that the U.S. tax system is not only a direct and intended source of revenue for the U.S. government, but it is also an indirect and unintended source of campaign funds for Congressional candidates. The link and unintended interaction is as follows: Congressional candidates have a direct stake in keeping the U.S. tax code complex and tax rates high for the wealthy. By promising to work for lowering tax rates for the wealthy, Congressional candidates have a never-ending source of campaign funds. Despite all the talk of flat taxes, candidates in both parties stand to lose greatly if it were actually enacted. No wonder that they are really opposed to the idea even though they can't say it because it would inflict great political damage. One cannot hope to understand, let alone reform, the tax code unless the *entire system* of campaign finance is understood and reformed.[8]

13 Keep timelines of different messes over time and how they interact and are "parts" of one another. For example, the education mess, the financial mess, and the health care mess are parts of one another and they interact. For instance, after a financial crisis, when the economy goes into a recession, those who lose their jobs (and health insurance) often want to go back to school to earn degrees

DOI: 10.1057/9781137386045

while waiting for the economy to recover. But the same financial crisis also forces schools to reduce the number of enrolled students and/or increase student fees because of school budget cuts. This is a typical example of how messes interact.

14 Bear in mind that every proposed "solution" becomes an integral part of the mess to which it is attempting to respond. Every proposed solution spans its own set of problems. Ideally, the new problems are "better" than the old ones in the sense that they are more easily resolved. The only way to assure this is to examine explicitly the consequences of different proposed solutions.

15 Examine carefully different "wild-card stakeholders." These are the "seemingly insignificant stakeholders" like Rosa Parks that spark a revolution (the Civil Rights Movement). Or Mohamed Bouazizi, who immolated himself thereby setting off the Tunisian Revolution. These are the ones who "can't and won't take 'it' anymore." Wild-card stakeholders may themselves seem insignificant but when the whole system/society is at the edge of chaos, even the most insignificant stakeholders, events, or interactions may trigger a chain reaction of events and interactions that can lead to major crises.

16 Make sure the least well-off stakeholders benefit. As an ethical rule of thumb, messes must be handled in such a way that the least well-off stakeholders benefit. Ask who are the known and unknown stakeholders that stand to gain the most/least? How will the most vulnerable versus the most well off fare? How will the poor and disadvantaged be affected?

Rules for Intervening/Presentation

17 Carefully Manage Presentations and the Degree of Challenge They Present

   a. Do Not Overwhelm One's Audiences: Increasing the diversity of perspectives and attempting to make sense of complexity can create high levels of anxiety and can thus be overwhelming. Thus, one needs to lead up to complicated power points and systems diagrams in carefully orchestrated steps, and not show the full diagrams all at once. The purpose of showing messes is not to confuse and overwhelm one's audience, but to help them understand and tolerate complexity. Nonetheless, there is no getting around the fact that the appreciation of messes and the ability to handle them requires a high tolerance for ambiguity.

DOI: 10.1057/9781137386045

If there are more than ten factors, which there always are, then prepare more than one diagram.

b. Rock the Boat (or let boat keep rocking—in a sense, maintain the status quo). When there are no better options left, create/let happen series of "minor" crises in the hope that crises will shock people to their senses. *A major, if not very risky, assumption is: a prolonged, sustained series of minor, contained and containable crises are the ONLY way to force people to abandon the status quo and to move off their deeply entrenched, divisive ideological positions. Another major assumption is that minor crises can be contained and not turn into major ones.*

From the standpoint of messes, the definition of a crisis is as follows: A major crisis occurs when the interactions that are seemingly the most invincible break down; a mega crisis occurs when a substantial majority of desired, planned interactions break down. Letting the boat rock will bring to surface and force us examine our faulty assumptions about improbable, insignificant, unimportant, easy/hard to manage, etc. interactions. Again, the danger is of course that "minor" crises can lead to "major" ones that can spin wildly out of control.

18 Pick Your Battles

a. Easy Wins: Go after the easiest to manage/understand interactions and by making headway on them build hope and show that it is possible to achieve change with and/or without revolution or major (mega) crises. For example, which teacher's unions in which states are most amenable to working with if not participating in the design of charters?

b. Magic: Court/Slay the Monster. Go after the most difficult to manage/ understand interactions and by making headway on them show that it is possible to achieve change with and/or without revolution or major (mega) crises. Again, which teacher's unions in which states are most amenable to working with if not participating in the design of charters?

c. In every complex situation, organization, institution, system, etc., there are always things (values, culture, rules, structures, friendships, pay and reward compensation, etc.) we would like to preserve or keep the Same (the status quo), and there are always things we would like to Change, sometimes radically. Conversely, there are always some things that are

DOI: 10.1057/9781137386045

Difficult versus Easy to Change or keep the Same. Thus, there are another four quadrants to consider with respect to organizational change.

Because of their very nature, messes have an abundance of issues in the Difficult to Change quadrant. This is precisely why transformative leadership is necessary in order to cope with messes. No wonder messes are difficult to manage for they call for heroic leadership, not managership.

19 Intervention Scale and Scope
   a. Use global/macro interventions. Foster Special Interest/World-Wide Groups for Taking Charge of/Managing Messes.
   b. Use grassroots interventions.

Managing heuristics

Nineteen heuristics is a formidable number to manage under any circumstances. Indeed, the nineteen heuristics are in danger of becoming their own mess! And in general, this is generally true. *We are always managing one by means of others.*

This is precisely why Figure 4.2 is the key to this whole book. Figure 4.2 is the centerpiece for monitoring the state of education and TEM. Unless we descend into complete chaos, we need to keep it clearly in mind.

Each of the various heuristics is a metric for assessing the state of Figure 4.2, e.g., how much it is realized and spreads beyond HCZ, what the greatest threats to it are, how they can be overcome, what new alliances are forming to assist its spread and further development, etc.

Finally, we cannot emphasize too much an earlier point. However difficult and demanding it is, if we didn't have an ideal model of any sort for not only assessing the current state of education, but for pointing how to move beyond it, then we would truly be in a hopeless mess.

Concluding remarks: the age of educational experimentation

We want to end this chapter on a strong recommendation. It is also a strong prescription that involves many of the preceding heuristics.

DOI: 10.1057/9781137386045

Universities need to establish new programs—entirely new colleges as it were—that integrate business, education, medicine, and social work, to mention only a few of the many academic disciplines and professional schools that need to be involved. In this way, students and practitioners will ideally be better prepared to confront and cope with TEM. Similar programs need to be established at the city, state, and federal levels of government.

Many years ago, the President of Tufts University suggested that the entire university become a University of Environmentalism. Every academic department, program, and research center would be involved in environmentalism in a significant way. Physicists and psychologists would still do physics and psychology, but in addition, they would work together to show how their disciplines contribute to our knowledge of the environment.

Unfortunately, the idea was never taken up. More than ever, it needs to be seriously considered and implemented. To say the least, we need at least one major university that is devoted entirely to K-12 education. We can no longer afford not to have at least one!

In this regard, HCZ needs to partner with a major university in order to become an integral part of an existing one. If no existing university will step up to the plate, then HCZ needs to consider seriously founding a college and/or university of its own so that it can "ramp up" and thereby spread its contribution to K-12 education.

In sum, the time is way overdue to acknowledge the existence and the importance of messes. We either learn to manage messes or they will continue to mismanage us.

Finally, as the epigraph to this chapter indicates, educational experiments (interactions) are exploding and will only continue to do so. Whatever the long-term fate of charters, the genie is out of the bottle. Education will never be the same.

Notes

1 Foroohar, Rana, "These Schools Mean Business," *Time*, April 9, 2012, p. 26.

2 Gilligan, James, *Why Some Politicians Are More Dangerous Than Others*, Polity, Cambridge, UK, 2011.

3 See http://www.hole-in-the-wall.com/Beginnings.html

4 Grubbs, F. E.: 1969, "Procedures for detecting outlying observations in samples," *Technometrics* Vol. 11, pp. 1–21.

DOI: 10.1057/9781137386045

5 Gladwell, Malcolm, *Outliers: The Story of Success*, Little, Brown and Company, Boston, 2008.
6 McKenna, Maryn, "Clean Sweep: Hospitals Bring Janitors to the Front Lines of Infection Control," *Scientific American*, September 2012, pp. 30–31.
7 Lessig, Lawrence, *Republic Lost*, Twelve, New York, 2011.
8 Ibid.

DOI: 10.1057/9781137386045

8

Waiting for Wilberforce—Making Sense of and Coping with the Tragic and Senseless

Abstract: *This chapter discusses school violence and shootings as messes from multiple perspectives. Indeed, they are integral "parts" of TEM. In particular, it discusses the general topic of school violence from each of the Jungian quadrants.*

Mitroff, Ian I., Hill, Lindan B., and Alpaslan, Can M. *Rethinking the Education Mess: A Systems Approach to Education Reform*. New York: Palgrave Macmillan, 2013.
DOI: 10.1057/9781137386045.

"... it is both shocking and predictable that James Yeager, the C.E.O. of a Tennessee company that trains civilians in weapons and tactical skills, posted a video online Wednesday (since removed but still viewable at rawstory.com) saying he was going to start killing people if gun control efforts moved forward. He said..."

"'I'm telling you that if that happens it's going to spark a civil war, and I'll be glad to fire the first shot. I'm not putting up with it. You shouldn't put up with it. And I need all you patriots to start thinking about what you're going to do, load your damn mags, make sure your rifle's clean, pack a backpack with some food in it and get ready to fight.'"[1]

"Threatened by long-term declining participation in shooting events, the firearms industry has poured millions of dollars into a broad campaign to ensure its future by getting guns into the hands of more, and younger, children."

"The industry's strategies include giving firearms, ammunition and cash to youth groups; weakening state restrictions on hunting by young children; marketing an affordable military-style rifle for 'junior shooters' and sponsoring semiautomatic-handgun competitions for youths; and developing a target-shooting video game that promotes brand-name weapons, with links to the Web pages of their makers."[2]

Introduction

As we indicated in the Preface, as this book was nearing its completion, the tragic and senseless shooting at the Sandy Hook Elementary School in Newtown Connecticut occurred. As a result, we not only felt we had to say something about school safety and security, but that we would be seriously remiss if we didn't.

But there is another important reason as well for treating school safety and security. Safety and security are not only integral parts of TEM, but they are major parts of it. As a result, they cannot be ignored. One cannot discuss TEM without discussing school safety and security at some point.

In addition, school safety and security constitute important messes in their own right. Thus, by discussing school safety, security, and violence,

DOI: 10.1057/9781137386045

we can apply, and thereby hopefully understand better, some of the heuristics of the previous chapter in greater detail.

The Jungian framework

Figure 8.1 shows the general plan and outline of the chapter. As we have done throughout, we basically use the Jungian framework to organize some of the many approaches to school safety, security, and violence.

The ST quadrant is concerned fundamentally with the basic definitions and measures of school safety and security and their dialectic opposite, school violence. It is also concerned with the profiles of those who commit school shootings and violence in general. It is thus basically concerned with what we know from cognitive and social psychology as well as sociology that allow us hopefully to prevent school violence from occurring in the first place and to deal with it more effectively once it has occurred. For example, can we identify through profile analysis the basic indicators/signs of those who are highly likely to commit violence? And, if we can, what specific steps can we take to intervene to prevent violence?

At some point, the ST quadrant also touches on specific proposals from various stakeholders such as the NRA, e.g., to have armed teachers in every classroom. To be as clear as possible, we are not only in complete disagreement with the NRA on this particular issue, but with its general philosophy. This does not necessarily mean that we oppose all forms of gun ownership and all types of guns per se. It does mean that we are strongly in favor of the strict control and complete elimination of military assault weapons in civilian hands. We are also in favor of much tighter gun control laws.

A recent article in *The Nation* put it bluntly: "The US has 5% of the world's population, but accounts for half of the world's firearms and 80% of the gun deaths in the 23 richest countries."[3] We kid ourselves—we are in mass denial—if we think a population of 300,000,000 plus can have on average one gun per person and manage the total number of guns in any society in a responsible way.

In a word, the NRA makes TEM and The School Safety/Security Mess (TSSSM) hopelessly worse by first of all not recognizing the existence and nature of messes—in short, by not thinking systemically—and as a result, secondly, by not dealing with TSSSM as a prime mess. If we have

DOI: 10.1057/9781137386045

learned anything about messes, they are not managed—coped with—by simple, black/white "solutions" such as increasing the numbers of guns in schools and in the general population at large. But then, the bitter failure to learn this crucial lesson is one of the key components of the mess.

The NT quadrant basically applies Systems Thinking to The School Safety/Security Mess. For instance, our ability to confront, let alone manage, school violence is affected by a great many diverse factors. For instance, (1) the tremendous number of guns that are easily obtainable in US society, (2) the prolonged and repeated exposure of children on a daily basis to seemingly endless amounts of media and movie violence, (3) the accessibility and proliferation of gruesome video games, (4) the increase in cyber-bullying and bullying in general, (5) the breakdown of traditional values and families, (6) lack of adequate parental supervision, etc. Because the factors are so numerous, can the relative contributions of each of these factors be assessed? Are some more important than others? Or, are the factors so intertwined that they cannot be assessed independently of one another? In other words, do they constitute a messy system in their own right?

Finally, the epitome of the NT approach is captured in various proposals to develop so-called smart guns that are extremely sensitive to the mental state of potential shooters and the context in which guns are deployed (e.g., schools, shopping malls, theatres, etc.) so that they would

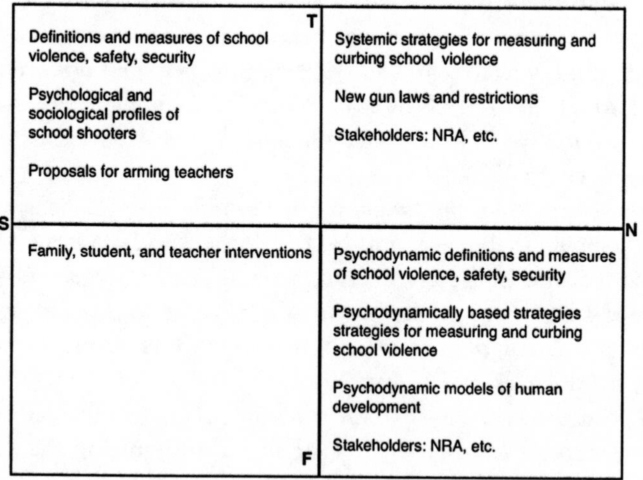

FIGURE 8.1 *Jung and school safety/security*

DOI: 10.1057/9781137386045

shut down automatically in the hands of the wrong person in the wrong circumstances.

Whereas the ST quadrant basically applies traditional psychology and sociology to uncover overt, more easily observed, and assessable factors to our understanding and our ability to cope with TSSSM, the NF quadrant digs deep beneath the surface of the forces that govern everyday life to examine unconscious, less easily available factors that are responsible for human violence. In particular, it examines the earliest phases of human development and the human propensity for violence.

Finally, the SF quadrant looks at how individuals cope emotionally with tragedies such as Sandy Hook. In particular, it discusses the admonition by rabid gun proponents to let emotions die down before proposing new gun regulations in the wake of Sandy Hook. Since the essence of SF is a deep emotional response to any and all events, in effect, the admonition is one of the deepest forms of insults that one could level against SF.

Most of all, the SF approach looks at how SF and NF working together can develop entirely new form of dealing with gun violence such as Twelve Step Programs for Gun Owners that are modeled on the Twelve Step Programs of Alcoholics Anonymous.

As in previous chapters, it is virtually impossible to keep the factors in one Jungian quadrant strictly apart, from interacting with, and spilling over to those in others. Indeed, there is a valid meaning and interpretation of each factor in terms of each quadrant. That is, each of the factors has a type of "meaning and existence" in each of the quadrants. For this reason, we are not overly concerned if the meanings and interpretations, plus the basic factors themselves, spill over. In fact, as we have been throughout, we are primarily concerned with the interactions and interconnections.

At this point, we trust that the reader will be able to follow the diverse Jungian meanings that can be attached to each and every factor.

The ST approach

We begin our discussion of the ST approach to the issue of school violence with a particular genre of research that is not only the heart of ST thinking, but is its very epitome. This is the fact that many who study school violence have long sought specific, detailed checklists that they

DOI: 10.1057/9781137386045

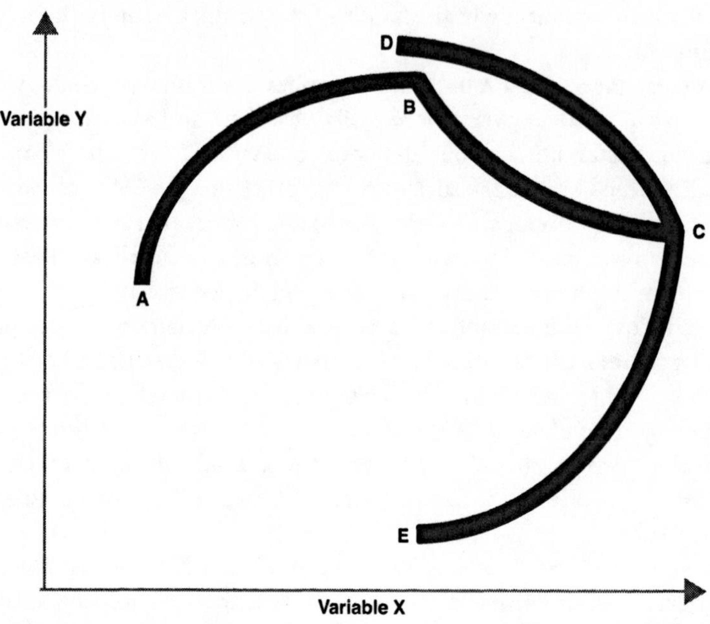

FIGURE 8.2 *Four patterns*

believe would not only identify those with a strong propensity to commit school violence, but with a high likelihood of their actually committing violence. We trust that by now it is abundantly clear to the reader why this is the essence of the ST approach. It is precise, detailed, and specific in its definition and assessment.

For instance, in *Books, Blackboards, and Bullets: School Shootings and Violence in America*, Marcel Lebrun presents a variety of checklists, as well as practical advice, for parents and school officials for preventing and responding to all forms of school violence including bullying, gangs, shootings, and suicide.[4] In "Tragedy and the Meaning of School Shootings," Bryan Warnick, Benjamin Johnson, and Samuel Rocha present a fair but balanced critical assessment of Lebrun's book.[5] First of all, Warnick, Johnson, and Rocha note that Lebrun has constructed his checklists from credible sources such as the FBI Threat Assessment Report. Thus, the basic, internal validity of the checklists is not in question. What is in question is the ability of parents, teachers, and school officials to use the checklists accurately and reliably, and from an even broader system's perspective, what their use implies for the general

DOI: 10.1057/9781137386045

culture and well being of schools no matter what the validity of the checklists.

We quote from Warnick, Johnson, and Rocha:

> The first checklist [of Lebrun] ... asks school officials to judge the frequency, duration, and intensity of a range of student personality characteristics. Educators are asked to judge, for example, whether a student has "difficulty with insults," whether they have an "inflated sense of self," whether they show "signs of narcissism" or have "an accurate sense of how others perceive them." Educators are asked if a student has "negative role models" or belongs to a "closed social group"... Lebrun claims, "If a child demonstrates more than 50 percent of these characteristics, you may have an at-risk individual on your hands"... With respect to the family checklist, educators need to determine whether there is a "difficult parent relationship," whether parents "model careless use of weapons," whether there are "limits in television watching," whether "parents give in to the child's demands," or whether "traditional family roles are reversed," among many other things... With respect to the larger social dynamic checklist, for example, educators are to know the students' outside interests, the groups they affiliate with, their stated references to other school shootings, and they "lack a reality check"...
>
> The checklists bring with them a certain vision of how educators should see students and how educators see themselves. First, notice how very difficult it would be for educators to make many of these judgments. For one thing, many of the student personality traits are too ambiguous, too context-dependent, or too difficult to measure. How does an educator decide what constitutes an "extremist" group or a "reality check"? How is an educator supposed to estimate the frequency, duration, and intensity, of students' display of "depression," "narcissism," or "vulnerability"? Even if one could settle on a shared understanding of such things, it would be difficult to make the appropriate observations. Further, this would all seem to imply that the educator needs deep and pervasive access to a student's inner life. They would need such information not only about the students themselves, but also about their families. Educators would be required to know if families carelessly handle firearms, for example, but it seems unlikely that any such information would be easily forthcoming.
>
> To gather all this information about students, and to conceptualize the bits of data as possible warning signs that precede an act of violence, demands that educators come to see students in a new light. *Educators would be required to adopt the gaze of a prison guard... Teachers, who usually lack a background in criminology, would now be required to view their students from a psychologically suspicious perspective...* [Italics ours][6]

DOI: 10.1057/9781137386045

Notice carefully that if merely assessing a student's propensity to commit violence calls for educators to *think* in effect like criminologists and prison guards, then actually requiring them to carry loaded guns, concealed or not, in classrooms requires them to *become* guards! To say that this would alter irreparably the general culture of schools is putting it mildly. This is precisely why initially great sounding, but simple-minded, black/white proposals are never as good as they first appear.

Unfortunately, it is not hard to imagine horrible consequences such as students getting caught in the crossfire of opposing shooters, getting access to teachers' guns, etc. Furthermore, if as in the case of Fort Hood, a deranged army psychiatrist was able to do considerable damage in a setting where guns are an everpresent feature, then what it would be like in schools?

The quotes above from Warnick et al. demonstrate as much as any single source both the strengths and weakness of the ST approach. The strength is that as much as possible it identifies accurately, clearly, and precisely the full range of factors that are known to be associated with school violence. The weakness is that it calls for teachers to measure and to implement what they or anyone else cannot easily do.

This doesn't mean that we should throw out completely all ST approaches. It means that one requires other psychological perspectives, e.g., NT/NF, to take the spirit of the ST approach and to complement by forming intuitive, global judgments. In other words, one has to go beyond the so-called precision of The Expert Agreement and Analytic Rationalist Inquiry Systems to others such as The Multi-Model ISs that allow one to form "informed judgments about the state of the whole." In sum, no psychological function or Inquiry System is complete or can function entirely by itself.

In short, there is nothing inherently wrong with exposing teachers to the checklists and giving them training in their use. The fundamental error consists in thinking that completely on their own, individual teachers can fill out, make sense of, and implement the checklists.

For example, if the burdens on individual teachers to fill out and assess individual students on checklists are enormous, which they are, then there is no reason why groups couldn't share this onerous task. But this means that one would have to administer a basic task that is ST in an NF manner.

There is no reason why individual teachers could not be responsible for particular items from the checklists to assess and then as a group

DOI: 10.1057/9781137386045

aggregate their individual observations into combined, overall assessments of particular students. While this might get around the problems of how teachers could assess students, it would not necessarily get around the atmosphere it would create in schools. Schools are supposed to be sanctuaries from guns and violence, not embody or perpetuate them.

We want to close this part of the chapter by making some further technical points that pertain to all ST approaches.

Let's say that we agree with defining the problem of school shootings as one of identifying shooters before they are able to commit violence. In terms of this definition of the problem, the solution thereby involves our developing a profile or set of profiles of likely shooters. A profile helps us to split people into two groups: those who fit the profile (i.e., shooters) and those that don't (i.e., non-shooters). Because profiles are never perfect, there are also two different kinds of statistical errors one can make: false positives and false negatives (see Table 8.1).

These two types of errors are costly in different ways. Type 1 errors falsely identify non-shooters as shooters. As a result, among other things, innocent students may unnecessarily be questioned or put under surveillance. Such activities also cost time and money.

Type 2 errors falsely identify shooters as non-shooters. As a result, innocent students may die.

Both types of errors can easily lead to crises, a subject which we discuss more fully in the next chapter.

Unfortunately, both errors are negatively correlated in the sense that decreasing the cost and/or likelihood of one generally increases the likelihood and/or cost of the other. For example, if one wants to decrease the number of unnecessary interrogations of innocent students, then one increases the likelihood of not catching a real shooter.

TABLE 8.1 *The two different kinds of statistical errors*

		Reality	
		Shooter	**Not A Shooter**
Prediction	**Shooter**	Success Outcome (Shooters correctly identified)	Type 1: False Positive (Non-shooters falsely identified as shooters)
	Not A Shooter	Type 2: False Negative (Shooters falsely identified as non-shooters)	Success Outcome (Non-shooters correctly identified)

DOI: 10.1057/9781137386045

In addition to the two types of statistical errors, advocates of defining the school-shooting problem as one of profiling potential shooters may also suffer from a number of cognitive biases. In the Oscar award winning movie, *Twelve Angry Men*, one of the angry and frustrated jurors yells at the character portrayed by Henry Fonda's, "I'm gonna kill you!" Of course, he doesn't mean it literally, and as a result, the incident is soon forgotten. But if the juror had actually tried to kill Fonda's character later on, then the jurors would have naturally thought of the prior incident as an early warning signal.

After horrific tragedies such as school shootings, people have a deep need to explain what happened. One has to make sense of the senseless. As a result, one is almost always able to find a number of early warning signals such as something the shooter said or texted to classmates, posted on a website, the way he dressed, etc. This, of course, can be a combination of two very well-known cognitive biases: hindsight (everything is clear 20/20 in hindsight) and confirmation (we look for evidence that confirms our beliefs about the world and ignore evidence to the contrary). Unfortunately, without the necessary training in profiling as well as Crisis Management (which we discuss in the next chapter), and the mindset that understands that school shootings issue are messes and not simple exercises, picking up early warning signals, interpreting them correctly, transmitting them to the right people, and taking the right kind of corrective actions are almost impossible to do in real time.

Of course, the discussion hinges on the central, critical assumption that the problem of school shootings is one of correctly identifying shooters before shooting, and that to do this, one needs to develop profiles. This gives rise to another error we need to consider.

Defining the problem as one of developing profiles of likely shooters can be a classic case of the Type 3 Error: "Solving the wrong problem precisely." This is especially the case when, as we have seen, the development of profiles of likely shooters that one can easily administer is extremely difficult to accomplish.[7] The possibility of a Type 3 error reminds us that we may not only be wasting time and money trying to solve the wrong problem, but we may even be making the original problem worse. For instance, the NRA has defined the problem of school shootings as one of stopping "a monster from killing our kids." Therefore, the natural "solution" is to have teachers or armed security guards at schools. The NRA itself has estimated that this would cost $2–3 billion a year. It could also create a very inhospitable environment

DOI: 10.1057/9781137386045

in which more shootings are likely to take place, and in this way "completely backfire," pun intended.

But something else makes the NRA's solution impractical. Most schools have first or ground floors with anywhere from 5 to 20 entry doors which are required for fire and safety reasons as well as multiple classrooms on the first floor. It would take a platoon of armed guards to defend each and every one of the schools.

Once again, problem definition is one of the most critical aspects of problem solving. Before, one can "solve" a problem, one first has to define it correctly. One then has to determine what form of disposition is appropriate, i.e., whether it is more important to resolve than it is to solve the problem, etc.

Finally, in an unexpected aspect that bears directly on one of the most prominent aspects of ST, namely finance, California, the nation's largest teachers' pension fund, voted to divest itself from companies that make guns and high-capacity ammunition magazines that are illegal in the state. The divestment process was begun after pension fund officials determined that it invested in a company that manufactured one of the types of weapons used in the Connecticut school shooting. The California State Teachers' Retirement System's investment committee unanimously approved a motion to divest.

The NT approach

There are many different types of NT approaches to violence. Since we can't cover every one, we are going to illustrate only two.

The first and most basic is the fact that the NT approach responds very differently to the idea that a great many factors are responsible for school violence. These include but are not restricted to the following: the constant and repeated exposure of children from an early age to TV, movie, and media violence; poverty; violent video games; the easy availability and access to guns; poor mental health programs; family violence; a weak economy; etc. All of these and more contribute to school shootings, bullying, etc.

The problem of course arises when one tries as the ST approach constantly does to determine the exact amount that each factor contributes individually and in combination with others to school violence. This is reflected in two widely differing positions (ST versus NT) in treating

DOI: 10.1057/9781137386045

gun violence that surface virtually every time there is a horrendous tragedy like that which happened in Newtown, Connecticut. Indeed, they are always just beneath the surface of any argument pertaining to gun control.

The first (ST) is represented by the movie/TV and video game industries, the second (NT), by cardiologists of all people.

The cardiological principle: attack every factor as aggressively as possible

The first position (ST) argues that there is no firm causal relationship between (1) the prolonged exposure of children and young adults to violent movies/TV/video games and (2) their engagement in actual violent behavior. Correlations are all there are, and correlations are not hard, definitive proof of causality, which of course is true. (As we have stressed throughout, this is even more true of messes where no one single factor in and of itself is completely responsible for the behavior of the mess.) Therefore, lacking such proof, there is no valid reason for the producers/writers of violent movies/TV/video games to tone down their creations. Besides, doesn't the First Amendment protect them?

The second position (NT) argues that no cardiologist would ever say that because a certain set of factors such as a high fat diet are low in their overall contribution to heart disease that one should therefore ignore them. Instead, no matter what their level of contribution, one should treat any and all factors as aggressively as one can. This is applicable to messes.

To draw out the differences between these two positions even more sharply, let us put them in the form of two opposing ethical principles because that's what they really are. The first says in effect that, "Whenever the correlation between what we do/produce as an industry and some important problem in society is low or beneath a certain 'threshold,' then we are warranted ethically in not doing anything; we are absolved as it were." The central question of course is, "How high would the correlation have to be before one accepted 'ethical responsibility'?"

The second says, "No matter how big or small the correlations, do everything in your power to make them even smaller."

The first principle is in effect a variant of Utilitarian Ethics. As such, it is a form of Cost/Benefit Thinking. One weighs the "benefits" versus

DOI: 10.1057/9781137386045

the "costs" of doing or not doing anything and if the benefits exceed the costs, or the costs are unclear or ill-defined from a ST perspective, then one is justified in engaging in a certain set of actions.

The second is Kantian in spirit. It says, "Make it a universal principle to do as little harm as possible." Even if the correlation is as low as 0.001 so act as to make it continually even lower. Indeed, Kant's theory of ethics is captured in his one of his most important dictums, namely, to treat all human beings as ends, and not as means. In effect, Cost/Benefit Thinking treats humans as means, not as ends.

As we have stated repeatedly, because gun violence is a complex mess, it cannot in principle be separated from mental health, video game violence, poverty, etc. But this doesn't mean that we shouldn't attack one of the most critical components of the mess, the availability of military style weapons that makes it far too easy for virtually anyone to possess or gain access to them.

In a supreme bit of irony, as the second chapter-opening quote shows, even though the gun industry has strongly condemned video games and has attempted to put the blame on them for causing violence, it is not adverse to using them to promote its products.

Few have put it better than Dennis Henigan in his book, *Lethal Logic*:

> ... Shouldn't Columbine have made keeping guns out of the hands of violent kids our most urgent national priority, while we also figure out how to deal with the far tougher challenge of the hardness of so many young hearts? No one can deny the importance of addressing the root causes of youth violence, but the need to do so is simply not a good reason to oppose laws to make it more difficult for kids to get guns ... [8]

While the correlations between viewing the worst depictions of violence may not rise to the level of "causality," over 35 years of research demonstrates that the correlations are not negligible. (A correlation of 1.0, the highest possible score, is not in itself proof of causality. The fact that two variables always occur in tandem is only part of the conditions that are needed to establish causality. One also needs to know that a certain variable X is sufficient to cause another variable Y, and that X is necessary for Y. Thus we need to know that if X occurs, i.e., precedes Y, then Y always occurs, and if X doesn't occur, then neither does Y.) That is, they are not zero. Typically, the correlations vary between 0.2 and 0.4, but even more significantly, they rise considerably with respect to children from high-risk environments, i.e., children living in poverty,

DOI: 10.1057/9781137386045

gang-infested neighborhoods, those who are left unsupervised for long periods of time, watch TV incessantly, etc. And, they rise with increased, persistent exposure to violence.

(The fact that recent research that shows that playing video games contributes to high levels of cognitive development is almost totally beside the point. The question is not only what playing violent games contributes to moral development, but even more to the point, is this the way as a society that we wish to raise the cognitive development of children? Is one really surprised to find that there are always some positive benefits in even the most odious things? But does this justify them?)

The arguments, which are integral parts of the mess that rabid gun proponents (but be it noted, not all responsible gun owners) use to justify the manufacture and possession of the worst possible types of weapons, are so many (e.g., "Gun Laws Are Generally Ineffective," "We Don't Need Anymore New Laws," etc.) are such that it is beyond the scope of this book to address them all. For a more complete and an excellent discussion, the reader is referred *Lethal Logic: Exploding The Myths That Paralyze American Gun Policy*.[9]

Instead, let us attack head-on one of the most flawed and central arguments that is used far too often: "Guns don't kill people; people kill people." This argument is not only "wrong," but it is "dead wrong." Pun intended! The more complete, correct rendering is "People kill people more effectively by means of guns than by any other known weapon to which they have easy access!" Yes, people can always kill people through the use of knives, clubs, poisons, etc., but does anyone seriously believe there would be the same amount of carnage if school shooters had instead used knives or clubs? Wouldn't they finally be overwhelmed and wrestled to the ground?

And yes, one can of course also kill people by means of cars, etc. But the fundamental point is that baseball bats and cars are not manufactured in the first place for the prime purpose of inflicting harm and killing. This is all the difference in the world.

Japan has very strict gun laws, and virtually no school shootings. There was one school "knifing" in 2001 by a former janitor who killed eight children and wounded several others. Of course, Japan also has very low crime rates,[10] and a very different culture that is much more collectivist than the US' individualist culture. Again the primary point is that guns are so much more deadly efficient than knives when it comes to killing people.

DOI: 10.1057/9781137386045

What nuclear weapons have to teach

The second form of the NT approach to school violence that we want to explore comes from of all places the study of nuclear weapons.[11]

In 1986, Mitroff published a study that he conducted of nuclear strategy. It showed that the numbers of paradoxes that were associated with nuclear weapons were far greater than previously recognized. Indeed, the whole phenomenon was riddled with paradoxes from beginning to end.

A central, main paradox was of course recognized almost from the very start of The Cold War between the US and the then Soviet Union: *Nuclear weapons existed for the prime purpose of NOT being used.* Both sides possessed great numbers of weapons that they could destroy the other many times over no matter who started a nuclear war. Indeed, both sides possessed enough nuclear powered submarines that they could remain submerged for months and thus hide the nuclear missiles aboard them in the vast oceans of the world. Therefore, if one nation was attacked, the other would still be able to retaliate with devastating consequences to both the initiator as well as the responder to a nuclear attack. It was hypothesized that a nuclear war would kick up enough dust to block out the Sun and plunge the entire Earth into a prolonged Nuclear Winter. No wonder that MAD, Mutually Assured Destruction, became an appropriate acronym for the situation in which both sides found themselves.

Nonetheless, Mitroff found that the whole issue of nuclear weapons was haunted by a host of even more thorny paradoxes. To see this, consider the following.

If one has missiles with nuclear warheads that are capable of striking enemy targets thousands of miles away, then one naturally wants to protect them as best as one can. One way to do this is to place one's missiles in underground silos that are protected by formidable amounts of concrete on top of them. In this way, they would be able to withstand direct hits from opposing enemy missiles.

The idea behind this strategy or type of thinking is best captured succinctly by the pithy expression, *More Leads To More*. Something initially thought to be good—more concrete—leads to more of a desired outcome, greater security in the form of enhanced protection of one's missiles. The trouble began when this seemingly sensible strategy led to its direct opposite, and hence, a major paradox.

(More Leads To More is one of the prime characteristics of the philosophy of The Machine Age, i.e., the Age of the Industrial Revolution where again everything was thought to be a machine. One of the major tenets of this way of thinking was, and still is, that if a certain amount of force, say X, was not sufficient to get a job done, then one keeps increasing the amount of force indefinitely, i.e., 2, 3, 4,... times X, etc.)

Putting more concrete on top of one's missile silos only encouraged the Russians to put multiple nuclear warheads on top of their missiles so that they could penetrate our silos and thus destroy our missiles. Thus, instead of More Leading To More, *More Leads Or Led To Less.*

Following this line of reasoning, it began to dawn on both sides that perhaps *Less Leads To More.* Putting less concrete on top of silos would threaten the US and Russians less and thus lower the chances of a nuclear war that neither side wanted. But then, there was still the sinking feeling that *Less Leads To Less.* Not doing everything one could to protect one's missiles left one in an extremely vulnerable state. Or, having fewer missiles than the enemy placed one in an extremely precarious position. In this way, the arms race endlessly fueled itself.

What Mitroff finally realized was that because of the inherent complexity and global scale of nuclear weapons, the phenomenon of nuclear weapons was in all four categories simultaneously! That is, *More Leads To More, More Leads To Less, Less Leads To More, and Less Leads To Less were all operating at the same time.* No wonder why it was virtually impossible to form a coherent nuclear strategy completely and solely from the standpoint of ST alone. For if anything seriously perturbs ST thinking with its presumption of nice, neat categories that pertain to relatively simple and stable phenomena, it is the existence of paradoxes. And, more the paradoxes, the worse the situation. Once again, More Leads To Less.

(What's interesting is that the same applies to all global phenomena such as the world economy. For instance, more complex financial instruments do not necessarily improve the world economy. Indeed, as we saw in 2008, it can throw it into a major recession.)

We hope by now that it is obvious how all of this pertains directly to guns. Indeed, it pertains to all phenomena especially in an increasingly tightly coupled and interactive world. (Again, global finance.) There is no question that at first—in the small—more guns do lead to greater safety, or at least a felt sense of safety and security. But, we believe that we have long ago reached a tipping point where More Does Not Necessarily Lead To More. Unfortunately, to understand and even more to accept this, one must be able to think systemically which unfortunately many are not able to do.

DOI: 10.1057/9781137386045

No schools, no shootings?

The NT approach is also known for its out-of-the-box thinking, and its extensive use of heuristics that examine extremes, never accepting conventional, traditional constraints or boundaries. Let us examine this aspect one step at a time.

Most schools are repositories of dense student populations in relatively small spaces, eight hours a day, five days a week. Moreover, those spaces are tightly controlled with very little flexibility for evasion and egress for students. There are two crucial variables here: 1) proximity—all schools have students grouped together every day in classrooms, auditoriums, dining areas, playgrounds, or athletic areas. (Irrespective of all lock-down and security measures, large numbers of students are still confined in small spaces almost continually); and, 2) student density—while no space is absolutely safe from mass trauma, schools with enrollments of 1000+ students simply exacerbate the density problem. Given this, is it possible to mitigate exposure of large numbers of students consistently available to the perilous intents of a mentally deranged shooter? One alternative may be through the continuing elaboration of effective learning technologies that render traditional classrooms/schools less necessary. Rick Ogston, CEO of Carpe Diem Schools of Tempe and Phoenix, Arizona and Indianapolis, Indiana, has developed a high-performance, high-reliability educational delivery system that could possibly reduce the proximity/density variables significantly. Through curricular and assessment advances via the Internet, high quality instruction can be delivered off-site to individual student computer stations for most of an instructional week. For example, if students are engaged in web-based instruction, such as the evolving "Digital Aristotle for Everyone" program, three days a week individually and offsite, with back to school to reviews and support two days a week, this reduces the proximity/density opportunity by 60%. And that is just the start. Of course, this alternative does not "solve" TEM in a single stroke because it gives rise to another host of countervailing concerns. If schools are in session for only two days per week, how do working parents provide for their children's care and education the other three days?

There are at least two more options that come into play, however unlikely or extreme they might be and therefore limit widespread acceptance and their eventual implementation. Education in the United States could move to a universal, home-school delivery system, thus requiring little or no aggregated student events. Lastly, the most radical

DOI: 10.1057/9781137386045

of the responses to proximity/density schooling is the notion of the de-schooling society in the tradition of Ivan Illich, albeit for different reasons, i.e., greater educational safety and security.

Nonetheless, whatever their form, the ST and NT approaches only take us so far. Something even more is called for: a deep appreciation of the NF meanings of safety and security, a topic to which we now turn.

The NF approach

The NF approach necessitates that we examine the deep underpinnings of human behavior. To do this, we turn to the pioneering work of some of the earliest child psychologists and psychoanalysts. Namely, we look briefly at the work of Melanie Klein, Donald Winnicott, and John Bowlby.[12] Each of them in their own profound way pushed back the boundaries of our understanding the earliest phases of the formation of the human psyche and personality, all of which starts from the very moment of birth. If this weren't back far enough, then the highly controversial psycho-historian Lloyd Demause claims to have developed a theory for the formation of personality by forces acting in the womb![13]

It is not necessary for the reader to accept fully the theories of the thinkers we discuss in this section to appreciate that NF meanings of safety and security differ markedly from the ST and NT meanings that we have encountered before. In accordance with an earlier point, the goal of the prior discussion was not to make teachers into criminologists or prison guards. In the same manner, the point of this section is not to make teachers into child psychiatrists or psychoanalysts. Instead, we believe that everyone who deals and works with children should have some familiarity with expanded notions of safety and security that are based on what we know from child psychoanalysis.

To this end, let us look briefly at some of the major insights from these early pioneers.

John Bowlby

What does the behavior of British children in WWII possibly have to do with guns and today's fractious politics? More than one would ever imagine!

DOI: 10.1057/9781137386045

In WWII, by being placed or lodged either in hospitals or massive care facilities, an overwhelming number of children were separated from their parents for weeks, months, and even years on end. As is well known, many were taken out of London altogether in order to protect them from the daily onslaught of German bombings. But worst of all were those who were permanently housed in orphanages.

When they first arrived at their new lodgings, the children cried for hours and days on end. This was especially true of those children who were housed in hospitals and orphanages. When they eventually stopped crying, they became zombie-like. They showed virtually no emotion whatsoever from that time on.

Attachment Theory arose out of the desire to understand better the horrific damage done to children that the British psychiatrist John Bowlby and others witnessed on a daily basis.

Bowlby and his colleagues found that two dimensions were key to explaining the emotional state of a child: Avoidance and Anxiety. Both were directly traceable to and the direct result of the emotional state of a child's primary caretakers. During Bowlby's time, the primary caretaker was of course the mother, if not throughout most of history. Whether the primary caretaker was either high or low on Avoidance and Anxiety had a tremendous effect on the child's emotional development. (Today, we know that all of a child's primary caretakers play a significant role, not just the mother. Thus, in the case of British children who were separated from their mothers, how they were responded to—and, in far too many cases, not responded to—and cared for by nurses made a significant difference in their behavior.)

By means of the mother's intense and frequent interactions—how she held, looked at, and attended to her child's cries and general discomfort—the mother subtly and not so subtly communicated her emotional state to her child. In short, she communicated how comfortable versus how anxious she was in fulfilling her role as a caretaker.

Since the interactions took place from the very moment of birth, they were largely preverbal and hence unconscious. In this way, the mother not only passed on, but influenced significantly the child's subsequent emotional development and state, most notably one's basic sense of trust and comfort with other people. Indeed, these early interactions laid down the basic "models of trust" that a person carried with him or her their whole life. A host of longitudinal studies have shown, in fact, that the effects last an entire lifetime unless of course a person has undergone significant therapy to reverse earlier conditions.

DOI: 10.1057/9781137386045

Since one can be either high or low on Avoidance and Anxiety, there are four primary combinations or states: 1. High Avoidance and High Anxiety; 2. High Avoidance and Low Anxiety; 3. Low Avoidance and High Anxiety; and, 4. Low Avoidance and Low Anxiety. State 1 is labeled Anxious-Avoiders; state 2, Avoiders; state 3, Anxious; and, state 4, Secure.

Those who are high on Avoidance exhibit, at least on the surface, little need or regard for other people. In the beginning of life, they were saddled by caretakers who showed little regard for them as a person. In short, their basic needs were met superficially at best. As a result, at a very early age, they gave up expecting anything from other people.

Those who are high on Anxiety were saddled by caretakers who, while he or she wanted to meet the basic needs of their child, experienced noticeable anxiety with regard to their capability of being able to do so. As a result, they are anxious around others because they are afraid they will either be abandoned or ignored. In short, they are needy.

Again, on the surface, Avoidants have little if any need of others and experience little if any anxiety in ignoring others. On the other hand, Anxious types want desperately to be around and to be liked by others but are terribly afraid that they won't. In a sense, Anxious types are perpetually striving to recapture the love of caretakers that showed, or were afraid to show, little emotion toward them even though they wanted to show positive emotions. Where Avoidants have basically given up, Anxious types are perpetually seeking to regain what they never had.

Where Anxious-Avoidants share the worst of both worlds, Secure types have the best. Secures not only want to be around others, but are comfortable in doing so.

Studies have shown that a significant number of top corporate and government executives are Avoidants. Avoidants radiate strength and fearlessness, the very qualities our culture admires in leaders.

It should come as no great surprise that political Conservatives generally exhibit many of the qualities of Avoidants and Liberal Progressives those of Anxious types. Indeed, Avoidants correspond most closely to George Lakoff's Stern Father, Liberal Progressives to Lakoff's Nurturing, if not Anxious, Mother.[14]

Attachment Theory sheds special light on the extreme views of the Right in general and the Tea Party in particular. Even though Nobel-prize winning economist Paul Krugman and others have argued cogently that the federal deficits are not the worst problem facing the US and other

DOI: 10.1057/9781137386045

European countries in the short run, and thus it would be better to incur more debt in order to get people back to work and pay down the debt when more people are able to pay taxes, debt horrifies the extreme Right. The very thought of, one, "being owned by others," whom of course they can never trust, and two, "others getting something for which they have not worked," goes against every grain of their psychic makeup. This is precisely why no set of logical facts or arguments will ever be enough to convince them otherwise.

But, by the same token, because deep down Liberal Progressives have a need to be loved, and believe that their values are universal and thus shared by everyone, they cannot get it through their heads that logical facts or arguments are never sufficient to sway anyone, including themselves.

This does not mean that everyone who opposes stricter gun controls is necessarily Anxious or Avoidant. It does mean that the core identity of Secure types is not heavily bound up with guns. It also means that they do not have an inherent fear and automatic distrust of government. In a word, they are not gripped by fear and paranoia.

In a word, Attachment Theory gives us very different notions of safety and security. Although they are obviously related, physical and psychological security are not the same. Indeed, one can have one without the other.

In short, the deepest senses of safety and security are formed at the earliest stages of life. And, they are constantly activated and reinforced by all of the parties with whom young children come in contact. This is precisely why having armed teachers in schools can have the opposite effect. Instead of providing more of a felt sense of security, it can lead to less. That is, from a deep psychological point of view, More Guns Can Easily Lead To Less Felt Security, not More.

Melanie Klein

It is said that if Freud discovered the child in the adult, then the British psychoanalyst Melanie Klein discovered the infant in the child. Working with children as young as six months and up to nine years, Klein discovered the earliest, formative roots of human behavior. Indeed, she literally invented play therapy as a way to understand what was going on in the minds of young children who were not yet mature enough to express in words what was going inside of them and their families.

DOI: 10.1057/9781137386045

Klein discovered that the earliest phase of child development was filled with extremely violent feelings and phantasies toward the child's principal caregivers. In the 1930s, the period in which Klein worked, this was again primarily the mother.

Klein aptly termed this phase, "paranoid-schizoid." It was "paranoid" because the child was literally terrified that the mother would abandon and/or harm it. It was "schizoid" because below the ages of 2 to 3, the child could not understand, and hence accept, that the "good mother"—the "good breast"—who met the child's every need at its beck and call was also the "bad mother"—"the bad breast"—that couldn't satisfy the child's every need at its every whim and fancy. The schizoid part was also called "splitting" because that's precisely what the child did in dividing the basic internal image of the mother into two distinct and irreconcilable parts or objects, i.e., the "good versus bad breasts."

(Splitting plays a prominent role in the thinking of the NRA because they assume tacitly that the world can be sharply divided up into good versus bad guys. In this simple-minded view of the world, good guys can have all the guns they want without harm to society as a whole.)

With patience, understanding, and tolerance on the part of the parents, and other key figures that cared for the child that splitting was a normal part of human development, the child eventually came to accept that the "good" and the "bad mother" were one and the same. In this way, the child not only healed the split he or she felt toward its mother, but the split within him or her as well. Just as important, the child came to accept that there were "good" and "bad" sides to everyone.

But, if for some reason the child was subject to trauma (beatings, sexual abuse, etc.), especially if it was prolonged and severe, then the split could become permanent. Splitting could also result if later in life one was subject to intense stress, disappointment, etc. Indeed, as adults, all of us from time to time split the world into "good" and "bad guys." Isn't this what unscrupulous politicians use from time to time to whip up support for their favored policies?

The key point is that if a person was not helped to heal splitting at whenever age it developed, then it not only festered and grew, but something even more ominous could result. The aggressive and violent parts of a person could be split off from normal moderating influences. In short, the aggressive and violent parts not only grew and intensified, but developed a separate life of their own.

DOI: 10.1057/9781137386045

In other words, splitting is more general than the division of people into sharply opposing images, i.e., good and bad guys. It also refers to the case where the acute aggressive impulses that are part of everyone are severed from moderating influences that are also a part of everyone.

Of course, none of this takes part in a social vacuum; what is going on in society and the general world around, one's work, family, etc. are also potential contributors to splitting as well. In addition, the aggressive and violent parts that could be split off from one another could be intensified further through association with other like-minded individuals to whom one was naturally attracted. Indeed, it was only with other like-minded individuals that one could express one's "true uninhibited feelings and emotions." Thus, over time, those with more moderate feelings would either be removed or remove themselves from such groups. In this way, groups would not only attract those who were extreme, but they would become increasingly more extreme over time.

This helps to explain why groups can often be more extreme than their individual members. The group first attracts, then amplifies those who share its basic beliefs, values, and viewpoints. In this way, groups can either intensify or ameliorate intense feelings of anger, contempt, fear, hate, paranoia, etc.

This also helps to explain the differences between those who have a deep versus a moderate attachment to guns. Those who grew up hunting and shooting guns at ranges as a natural part of life are naturally reluctant to give up an activity they enjoy and at which they are good. As difficult as these persons find it to give up guns for the greater good of society, it pales in comparison with those who are deeply gripped by fear and paranoia. In the later case, giving up guns is not just giving up a cherished part of one's basic identity, but it is akin to giving up one's lifeline, one's fundamental sense of psyche security.

Donald Winnicott

As richly deserving as it is, this is not the place to give an extended account of Winnicott's work. Therefore by necessity, we shall limit ourselves to merely two of his most important and major contributions: (1) the "mother-child" as a single indivisible unit and (2) the concept of the "holding environment."

DOI: 10.1057/9781137386045

As a pediatrician, Winnicott was resolute in his insistence that in the very early beginnings of an infant's life, there was no separate, distinguishable mother and child. Instead, from his extensive observations, the mother-child formed and had to form an indivisible, indistinguishable bond and thereby unit. In short, there was no mother or child, one of the prime statements for which Winnicott is justly famous.

The concept of the holding environment paralleled and reinforced the notion of the mother-child as an indistinguishable unit. The "holding environment" signified that in the beginning the mother was the "container" for all of the often all-too-intense emotions and feelings of the infant, most of which it couldn't contain or make sense of on its own. The mother thus not only held the baby physically, but just as important, she "held" the baby's feelings and emotions in check.

The concept of the holding environment also signified that the mother protected the child from internal and external threats.

Notice carefully how this applies to very young children. Schools are figuratively and literally "holding environments." This is precisely why having armed teachers goes deeply against the grain of schools as "psychological havens or holding environments." Notice also that in advocating more guns as the solution to gun violence, this ignores the fact that the "general holding environment" of society in the form of the interactions between restrictions on guns, proper mental health care, proper family care, curbs on violent video/movies/TV, etc. have all broken down or been seriously eroded. In other words, the "containers" of the "gun mess" have broken down.

The SF approach

As before, the essence of the SF approach is that of mourning and honoring the particular victims of school shootings. As such, the admonition of the NRA and rabid gun owners to "let emotions simmer down before considering responses to Sandy Hook" not only completely misses the mark, but is the worst insult one can hurl at SFs. SF deliberatively and instinctively brings pictures and mementos of the victims to the sites where they died. It wants to keep the memory of those whose lives were taken so senselessly alive for as long as possible.

In this regard, SF often pairs up with NF in forging deep links with the survivors of other tragic shootings. In this way, NF aids by proposing

DOI: 10.1057/9781137386045

general strategies and founding movements such as new web sites and organizations dedicated to gun control. Here is one example of a SF/NF solution to the school shootings problem:

> What if, for example, forgiveness mingled with sorrow for the shooters themselves was suggested, or was somehow made more visible, through ritual? Instead of being remembered with fear and awe, future shooters would know that they would be remembered differently: not as "badasses," but simply as badly treated children; not as wielding fearful power, but simply as being driven by powers beyond their control; not as gods of violence, but simply as young people to be pitied and forgiven. Could such forgiveness be publicly embodied in ritual? Could the community offer another ritual solution to the problem of the ritual inquiry? Would such ceremonial meanings preempt the script that troubled youths are trying to write?[15]

We want to explore what is to our knowledge is a novel way of incorporating SF and NF. If as we believe that an obsessive need for guns is akin to an addiction and therefore cannot be dealt with by means of conventional arguments (after all, many alcoholics know "rationally" that alcohol is killing them but they are still unable to resist it), then we believe that we need to stop beating around the bush and treat it as a major form of addiction. Accordingly, we have taken the Twelve Steps of Alcoholics Anonymous listed directly below and reworded them to apply to our society's obsession with guns. In suggesting this, we are not kidding ourselves that this in and of itself will help us to better manage what we believe is our society's completely out-of-control proliferation of guns.

The twelve steps of Alcoholics Anonymous

1. We admitted we were powerless over alcohol—that our lives had become unmanageable.
2. Came to believe that a Power greater than ourselves could restore us to sanity.
3. Made a decision to turn our will and our lives over to the care of God as we understood Him.
4. Made a searching and fearless moral inventory of ourselves.
5. Admitted to God, to ourselves, and to another human being the exact nature of our wrongs.
6. Were entirely ready to have God remove all these defects of character.
7. Humbly asked Him to remove our shortcomings.

DOI: 10.1057/9781137386045

8 Made a list of all persons we had harmed, and became willing to make amends to them all.
9 Made direct amends to such people wherever possible, except when to do so would injure them or others.
10 Continued to take personal inventory and when we were wrong promptly admitted it.
11 Sought through prayer and meditation to improve our conscious contact with God, as we understood Him, praying only for knowledge of His will for us and the power to carry that out.
12 Having had a spiritual awakening as the result of these Steps, we tried to carry this message to alcoholics, and to practice these principles in all our affairs. (Copyright _ A.A. World Services, Inc.)

THE TWELVE STEPS OF GUNS ANONYMOUS

1 We admitted we were powerless over our fascination with and need for guns and as a result that our lives and society as a whole had become unmanageable. (Note that this first step is an admission that one was in denial about guns.)
2 We came to believe that a Moral Power greater than ourselves could restore us to sanity. That is, the lives and well being of children were more important than our desire to hunt, shoot, and collect/own firearms, especially high-power automatic weapons. As such, we accepted that no rights were absolute. Thus, while we still believed in the Second Amendment, we came to realize that it did not sanction the possession of weapons of war.
3 We made a conscious decision to turn our will and our lives over to the care of this Moral Power as we understood It.
4 Made a searching and fearless moral inventory of ourselves and how our unrestrained possession of firearms harmed the collective good of society.
5 Admitted to a Higher Power however we conceived of Him/Her/It, to ourselves, and to another human being the exact nature of our wrongs.
6 Were entirely ready to have our Higher Power remove all these defects of character.
7 Humbly asked our Higher Power to remove our shortcomings.
8 Made a list of all persons we had harmed through our beliefs and actions and became willing to make amends to them all.

DOI: 10.1057/9781137386045

9 Made direct amends to such people wherever possible, except when to do so would injure them or others.
10 Continued to take personal inventory and when we were wrong promptly admitted it.
11 Sought through prayer and meditation to improve our conscious contact with our Higher Power, as we understood Him/Her/It, praying only for knowledge of His/Her/Its will for us and the power to carry it out.
12 Having had a spiritual awakening as the result of these Steps, we tried to carry this message to gun owners, and to practice these principles in all our affairs. In particular, we saw the need to design and implement a new organization of Responsible Gun Owners for the Collective Good.

In short, the practitioners of this new Twelve Step Program for Gun Owners would be enacting a new form of a Precautionary Principle for Children. That is, if there was the slightest chance that a specific type of weapon posed an especially dangerous threat to the well being of children in particular, then they would willingly give that weapon up for the greater good of society.

(Of course, how far to push the analogy with AA is a matter for future discussion. That is, should members attend group meetings as do the members of AA? We believe so, but again this is up for debate.)

As much as the SF Approach wants to mourn and honor the particular victims of school shootings, it often values the extremely hard but humane attempt to make sense of what it must have been like to be a shooter. What did they feel? Did they feel pain? Did they feel sorry? Why did they do it? In "Why Kids Kill", Peter Langman opens a window into the tormented souls of shooters. He has identified three different types of school shooters: traumatized, psychotic, and psychopathic. In other words, different types of shooters can no more be lumped together than can all different types of guns owners.

Traumatized shooters act out in the worst possible way the pent up reactions to their being victimized (sexually abused, physically beaten, tortured, humiliated, bullied, etc.) often repeatedly and over long periods of time by parents, siblings, teachers, fellow students, relatives, strangers, etc. Other than feeling existential rage, traumatized shooters are emotionally numb; they don't have enough capacity to feel close to others.[16] In short, the desire is to hurt as many others as possible in payback for all the ways in which others have hurt them repeatedly for

DOI: 10.1057/9781137386045

years and without anyone coming to their aid. In a word, they have been deeply betrayed by those who were supposed to take care of them and treat them with respect and dignity.

Psychotic shooters are basically out of touch with reality, whether through hearing strange voices and seeing strange visions. Their hallucinations make it impossible for them to feel empathy. As such, they are also impervious to argument and reason. Obviously, not all psychotics are shooters. Langman found that psychotic shooters are often young males who have significant conflict with their parents; these kids also fail to take their anti-psychotic medication, and suffer from substance abuse.[17]

Finally, psychopathic shooters lack the ordinary moral compasses that most people have. They are often habitual liars and charmers taking extreme pride in the fact that they can pull the proverbial wool over the eyes of most people. More often than not, their psychopathy starts early in childhood with the killing and torturing of small animals and pets.

A major characteristic of psychopaths is that they do not experience guilt and remorse in the ways, if at all, that ordinary people do. But since they have developed the ability to charm others a high degree, they are able to fake feelings and emotions to a high degree as well.

Obviously, not all psychopaths and sadists are shooters. Since sadism is one of the marks of psychopathology, when they are found in male adolescents who have a love of guns and have managed to convince a non-psychopathic partner to follow them, the results are lethal.[18]

Consider how each of these types is likely to respond to the naive assertion of Wayne Lapierre, Executive Vice President of NRA, that "The only thing that will stop a bad guy with a gun is a good guy with a gun." First of all, leaving aside the simple-minded, black/white "splitting" between so-called good and bad guys, traumatized victims are not stupid. They can assess reality to a degree, but if their trauma is so great, then they want to take out as many people as possible and they want to be killed (as part of a death wish) in order to end their misery. In this sense, deterrence per se will not necessarily be a deterrent as it would be to the ordinary person.

Psychotics are also not deterred because of their basic inability to reason in accordance with ordinary standards.

And, psychopaths are likely to take threats of deterrence as a challenge to "beat the 'marks' and in order to show them that I'm so much smarter than they are."

DOI: 10.1057/9781137386045

None of this is to say that teachers shouldn't be trained in how to protect themselves and the children under their care physically through such means as martial arts and other forms of self-defense such as barricading classroom doors, school doors, etc. Of course, they should. But we shouldn't turn schools into fortresses or kid ourselves that arming teachers is the answer.

Religious-like aspects

There is no denying that there is a deep, fundamentalist, if not cult-like, aspect to the makeup of many ardent gun owners. This is perhaps the strangest aspect to the whole issue for the founding fathers did not intend via The Bill of Rights to aid and abet any kind of "state religion." And yet, the fervor in which many hold that The Second Amendment is akin to an article of religious faith.

Every so often, a truly great book comes along that helps to explain why fundamentalism exerts such a hold on so many minds. *When God Talks Back: Understanding the Evangelical Relationship with God* by Stanford anthropologist T.M. Luhrman is certainly one of these.

Let us recount briefly the main argument of Luhrman's book. Doing so not only helps us to understand better the Christian evangelical mind, but strangely enough, why Democrats and Republicans are divided so strongly. In a word, Luhrman gives us deep insight into the nature of various belief systems and why the battle between them is often so bitter and prolonged. Thus, even though it seems far removed from the gun debate, it helps to illuminate one of its thorniest aspects.

People are drawn to Christian evangelicalism for a variety of reasons. Among the more typical are life-long struggles with addiction, alcoholism, a history of bad relationships, loneliness, social isolation, and the general feeling that they are missing something deep and fundamental in life. Accompanying these is also the feeling that one not only needs, but is ready to forge a personal relationship with God.

In terms of Inquiry Systems, the initial reasons are as we have seen "initial inputs or starting beliefs." More fundamentally, they are "tentative 'truths'" that the system will "operate on" in highly specified ways so as to produce a "final state of truth" or more generally "state of Being." This "final state" is typically not an abstract proposition but a strong prescription to engage in actions of some kind to change either oneself

DOI: 10.1057/9781137386045

and/or the world. The "final truth or state" we identified earlier as the "output" of a belief system. Since it is generally regarded as "established beyond doubt," it is therefore regarded as "The Truth."

Luhrman shows in great detail what the "operators" are in Christian evangelicalism that "transform" the "inputs" into "final established Truth." In short, they are a carefully orchestrated and prolonged series of "special spiritual exercises" such as distinct types of praying that train the mind first to imagine and then experience via all the senses a different reality.

The "output" is a "direct, personal experience and a day-to-day, ongoing, permanent relationship with God!" One no longer just "believes in God" in the abstract but "knows God intimately" as one would a personal friend. The derived benefits, and thereby additional "outputs," are enhanced calm and peace. The ultimate end is a "heightened emotional state."

As we also saw earlier, one of the most important components of a belief or Inquiry System is the Guarantor. The Guarantor is the set of underlying beliefs that are accepted without question. They are undeniably true. As such, they constitute the absolute, foundational bedrock of the entire system. In Christian evangelicalism, the Guarantor is the unquestioned belief that The Bible is literally true and that God exists without doubt. To say that The Second Amendment plays a corresponding role for fervent gun advocates is not stretching it at all.

Dennis Henigan has said it best:

> As one NRA leader put it some years ago, "You would get a far better understanding [of the extreme fervor with which gun owners often have for guns] if you approached us as if you were approaching one of the great religions of the world." This is not a frivolous comparison. There is an unquestionably religious fervor about the beliefs of many pro-gun partisans. It is grounded in various articles of religious faith that form the catechism of the NRA: that law-abiding citizens are under constant risk of attack by predatory criminals, that the safety of every person and family depends upon the ability of individuals to defend themselves with firearms, that the government cannot be trusted to provide security to individuals and families, that democratic institutions cannot be counted on to protect our liberties as Americans, that those institutions are at constant risk of subversion by tyrannical elements, and that tyranny is kept at bay only by the potential for insurrection by an armed populace intent on maintaining liberty. In the NRA's world, these are eternal truths.

DOI: 10.1057/9781137386045

> To the true believers, the gun is an object of religious devotion... The hallowed place of the gun is reflected in the holy text of the gun rights movement, the Second Amendment to the Constitution...[19]

As we read Luhrman, we thought constantly of the recent, drawn-out, and bitter debates between the candidates for the 2012 Republican presidential nomination. Although both parties constantly use emotional appeals, we believe that it is not an exaggeration to say that with its extreme tilt to the Right, the Republican Party is much closer to a Christian evangelical mindset than the Democrats. For example, we are still struck by the extreme emotional belief systems of Speaker Gingrich and Senator Santorum, not to mention soon-to-be former Congresswoman Michele Bachmann.

In effect, nearly everything in their so-called arguments is pure "output." That is, the sheer outrageousness of a claim, i.e., the "output Truth," is simultaneously the "input," "operator," and the "Guarantor." No wonder why Liberals and Progressives who believe so deeply in Reason are so offended and feel a deep sense of revulsion. Everything is not only hopelessly confounded and entangled, but sheer emotional dribble. There are no independent corroborating facts as it were.

Nonetheless, we would strongly caution Liberals and Progressive not to eschew emotion altogether. The proper moral of the story is that more than ever, Reason and Emotion need to work together. So-called logically pure belief systems may move scientists, but they are hardly sufficient to move the larger body of people to think great thoughts and/or to undertake great actions. Luhrman shows that at present Conservatives understand this far better than Liberals and Progressives.

Liberals and Progressives are not as smart as they are and we would like to believe they are. They have a lot to learn about belief systems before they can ever hope to change them.

Can progressives and conservatives agree when children are at stake?

In 2012 alone, there were 25+ school related shootings, ranging from near attempts to the horrific event at Sandy Hook Elementary. While the 1999 shootings in Columbine brought the tragedy of school shootings to the forefront, there has been both an increase in prevention measures, and yet sadly, similar shootings. For this reason alone, it is not only

DOI: 10.1057/9781137386045

informative, but imperative that we consider at least two overarching public policy variables related to school quality and school safety.

1 *The Social and Emotional Status of Children in the United States.* Mental health disorders—nearly all school shooters have exhibited significant psychological disorders that have attracted the attention of relatives, acquaintances, classmates, and school personnel. Despite this, fear, reluctance, and a host of privacy issues hamper preventative intervention. Hyper-traumatic incidents such as Columbine, Virginia Tech, Aurora Colorado theater, Sandy Hook Elementary School, and others exist side by side with societal concern for individual autonomy and freedom on the one hand and mass homicidal events that tear at the collective psyche on the other. Public policy solutions are complicated and elusive. Can therefore *both progressives and conservatives* craft an interventionist yet compassionate protocol for supporting and treating citizens with significant and potentially lethal behavioral tendencies? Can society maintain that delicate balance between individual autonomy and freedom with the need for safety and sense of community? *In short, can progressives and conservatives agree when all of our children are at stake?*
2 *High Capacity Weapons Availability*. Weapons with the capability of high-volume delivery were designed originally for war or wartime situations where intentional killing has been deemed necessary and acceptable by society. Historically, such weapons have been closely regulated and legally unavailable to all citizens, with the exception of certain military and law enforcement personnel. More recently, such weapons, although still legally unavailable to the general citizenry, were becoming available to criminals, and most recently, have become largely available to everyone. Progressive and conservative arguments for and against general gun control notwithstanding, it is a fact that certain types of weapons with very high-volume delivery capabilities are available almost universally. The very sad fact is that these types of weapons in the possession of (mentally deranged individuals bent on homicide, combined with high density, close proximity student populations have produced horrific, high-fatality tragedies. From a purely analytical perspective, the combination of these three variables has created a perfect storm. These three highly salient variables—homicidal mental illnesses, the high density/close proximity of student

DOI: 10.1057/9781137386045

populations, and high capacity weapons availability—are in a lethal combination just waiting for the next perfect storm. Despite incredible complexities of public policy implementation and the possibility of unintended consequences, is this really the best we can do as a society? *Once again, can progressives and conservatives agree when our children are at stake?*

Concluding remarks

In this chapter, we have not only tried to address the phenomenon of school shootings as a mess, but as such, to illustrate many of the heuristics of the preceding chapter.

First of all, we cannot say too much that the problem(s) of school shootings in the US is a mess. For this reason, we have deliberately sought the analyses of those experts who cross-connect fields. We have swept in ideas from a variety of disciplines and topics: from the NRA's armed teachers proposal to profiling checklists, from attachment theory to US politics, and even from Evangelism to theories of personality formation in the womb. We asked, "What can we learn from Cardiologists and Alcoholics Anonymous?" We wanted to cover topics as diverse as possible to even begin make sense of the senseless.

Using the Jungian Framework, we have addressed the mess from multiple perspectives in the fervent hope that we have not only drawn out important ideas regarding school shootings and gun control, but that we have illuminated new/atypical ideas/interactions as well.

We examined and challenged taken-for-granted assumptions and beliefs such as the belief that guns don't kill people, people kill people. We used psychoanalytic thought to examine the deep and thereby often unconscious assumptions that we make about different stakeholders. To make our points clear, we visited and examined extremes. We rocked the boat. If the reader did not feel uncomfortable while thinking through some parts of this chapter, we have not accomplished our purpose.

We wanted to ask smart/dumb questions, often playing the devil's advocate. The natural dialectic and conflict between quadrants helped us achieve this goal. As we will see in the next chapter, once again the various quadrants need and keep in check each other.

We paid attention to outliers such as the extremely low crime rate in Japan where no school shootings take place. We also paid attention to

DOI: 10.1057/9781137386045

counter-intuitive, paradoxical, and unintended interactions and relationships such as the lessons we learned from the Nuclear Arms Race of the Cold War era.

We did not intend in this chapter to offer permanent solutions to the School Shootings Mess. We have tried our best to ask the right questions.

Waiting for Wilberforce

Lastly, we want to end this chapter by calling attention to one of the greatest persons—an outstanding "moralist" in the best sense of the term—who ever lived, the Englishman, William Wilberforce.[20] Born in 1759, Wilberforce fought his whole life for the elimination of British slavery. He worked tirelessly to make the horrific conditions of slaves palpable to the British public. And, as a gifted orator, he spoke tirelessly in the British Parliament against slavery. At the very end of his life, he had the joy of seeing the end of British slavery.

We are waiting for our American Wilberforce to curb our national obsession with guns. Hopefully, we are not waiting in vain.

Notes

1 Blow, Charles,, "Revolutionary Language," *The New York Times*, Saturday, January 12, 2013, p. A17.
2 McIntire, Mike, "Selling a New Generation on Guns," *The New York Times*, Sunday, January 27, 2013, p. 1.
3 "Noted," *The Nation*, January 7/14, 2013, p. 5.
4 Lebrun, Marcel, Books, *Blackboards, and Bullets: School Shootings and Violence in America*, Rowland & Littlefield Education, Lanham, MD, 2009.
5 Warnick, Bryan, Johnson, Benjamin, and Rocha, Samuel, "Tragedy and the Meaning of School Shootings," *Educational Theory*, Vol. 60, No. 3, 2010, pp. 371–390.
6 Ibid, pp. 386–387.
7 Heck, W. P., 2001, The School Shooter: One Community's Response. *FBI Law Enforcement Bulletin*, Vol. 70, No. 9, pp. 9–13.
8 Henigan, Dennis, *Lethal Logic: Exploding The Myths That Paralyze American Gun Policy*, Potomac Books, Washington, DC, 2009, p. 35.
9 Ibid.

DOI: 10.1057/9781137386045

10 McMillan, Charles, *The Japanese Industrial System: De Gruyter Studies in Organization*, 3rd edn, p. .46.

11 Mitroff, Ian I. "The Complete and Utter Failure of Traditional Thinking in Comprehending the Nuclear Predicament, Why It's Impossible to Formulate A Paradox Free Theory of Nuclear Policy," *Technological Forecasting and Social Change*, Vol. 29, 1986, pp. 51–72.

12 See Grossman, Klaus, Grossmann, Karin, and Waters, Everett (eds), *Attachment from Infancy to Adulthood; The Major Longitudinal Studies*, Guilford, New York, 2005; Karen, Robert, *Becoming Attached: Unfolding the Mystery of the Infant-Mother Bond and Its Impact on Later Life*, Wraner Books, New York, 1994; Klein, Melanie, *The Psychoanalysis of Children*, Nabu Public Domain Reprints, Grove Press, New York, 1960; Spillhaus, Elizabeth, and O'Shaughnessy, Edna (eds), *Projective Identification: The Fate of a Concept*, Routledge, London, 2012; Winnicott, Clare, Shepherd, Ray, and Davis, Madeleine (eds), *Psychoanalytic Explorations*, Winnicott, D.W., Karnac, London, 1989.

13 Demausse, Llyod, *Foundations of Psychohistory*, Creative Roots, Inc., New York, 1982.

14 Lakoff, George, *Moral Politics*, University of Chicago, 1996.

15 Warnick, Johnson, and Rocha, op cit.

16 Langman, Peter, *Why Kids Kill*, Palgrave Macmillan, 2010.

17 Ibid.

18 Ibid.

19 Henigan, op cit. p. 77.

20 Metaxas, Eric, *Amazing Grace: William Wilberforce and the Heroic Campaign to End Slavery*, Harper Row, New York, 2007.

DOI: 10.1057/9781137386045

9

Crisis Management—An Imperative For Schools

Abstract: *This chapter discusses various aspects of Crisis Management in the context of schools. Indeed, because messes always have a high potential for crises, Crisis Management is thereby an integral component of our ability to manage TEM.*

Mitroff, Ian I., Hill, Lindan B., and Alpaslan, Can M. *Rethinking the Education Mess: A Systems Approach to Education Reform*. New York: Palgrave Macmillan, 2013.
DOI: 10.1057/9781137386045.

DOI: 10.1057/9781137386045

Introduction

The previous chapter represented our attempt to make sense of the school violence mess. As we noted, it was prompted because of the tragic shootings at the Sandy Hook Elementary School in Newtown Connecticut. But shootings or similar acts of violence are not the only crises that can strike schools.

While all crises are inherently complex and therefore messes, not all messes are necessarily crises. Nonetheless, precisely because of their messiness, i.e., complexity as well as high levels of uncertainty and ambiguity, messes always contain a high potential for crises. Another way to put it is that some crisis, however small, is always brewing in a mess.

An imprecise definition of crisis

The following are some of the extreme events that come to mind when the word crisis is mentioned: Johnson & Johnson's Tylenol poisonings, the gas leak disaster in Bhopal, NASA Space Shuttle accidents, Chernobyl, Mad Cow disease, the Exxon-Valdez and BP oil spills, the Oklahoma City bombing, airplane crashes such as ValuJet Flight 592, TWA Flight 80, and Air France Flight 447, the Ford-Firestone tire crisis, 9/11 terrorist attacks, the Internet/Nasdaq bubble and bust, Enron's collapse, WorldCom's bankruptcy, SARS, the East Coast power outages, Hurricane Katrina, the ongoing financial and economic crisis, the Fort Knox shootings, the Haitian and Chilean earthquakes, the Japanese nuclear power plant disaster, Penn State child sex abuse scandal, the child sex abuses in the Catholic Church, and more recently the Sandy Hook School Shooting. Obviously, there are many, many more that we could have included in the list above.

A crisis is a set of extreme events that threaten the viability of an organization. Moreover, they cannot be contained within the walls of the organization, and they always challenge, if not shatter, the basic assumptions of the stakeholders of the organization. As we have mentioned above, crises are inherently messy. They are "characterized by ambiguity of cause, effect, and means of resolution".[1] Thus, there is no one definition on which all stakeholders, who affect or are affected by the crisis, agree. In other words, what is a "crisis" and who is "stakeholder" is always in the eye of the stakeholder.[2]

DOI: 10.1057/9781137386045

Crisis management: an institutional imperative

Since the Tylenol scare in 1982, corporate leaders have become increasingly aware of the need to develop and implement Crisis Management programs. The general expectation is that the senior management of an organization will have effective Crisis Management plans in place to deal with a broad range of crises. Indeed, the literature on Crisis Management suggests that (a) Crisis Prepared companies experience significantly fewer crises than Crisis Prone companies; (b) Crisis Prepared organizations recover faster from crises; and (c) Crisis Prone organizations are significantly less profitable than Crisis Prepared organizations.[3]

Schools are not different from their corporate counterparts in the sense that they can be struck by a broad range of crises. The Sandy Hook High School Shootings are just one example of the kind of crisis that can strike an educational institution. There are a variety of others such as fires, explosions, suicides, unfounded rumors, natural disasters, child sex abuse scandals, food poisonings, grade tampering, and so forth. Schools are certainly not immune to the ups and downs of the economy. As a result of the ongoing global financial crisis, many schools have seen their budgets cut, the contributions of alumni and parents decrease, community support decline, etc. As a consequence, many institutions have increased tuition, implemented furloughs, laid off faculty and teachers, and postponed maintenance and projects.

School stakeholders, which in the large is society as a whole, but especially parents, surrounding communities, and of course, children, rightly expect that institutions will be well prepared for a broad range of crises beyond natural disasters, fires, and explosions. It is incumbent on the administrators, teachers, parents, and other employees of schools to learn from their corporate counterparts and to take an active role in insuring that their institutions are well prepared for a broad range of crises.

In this chapter, we focus on Crisis Management in schools and the role of school leadership in developing and implementing Crisis Management plans and programs. After we discuss proactive versus reactive Crisis Management approaches, the importance of crisis typologies, Crisis Management Teams, methods to enhance collaboration, the role of the news media, early warning signal detection, and institutional cultures in escalating or containing a crisis, we present recommendations that

DOI: 10.1057/9781137386045

school leaders can use to assess the capabilities of their institutions to respond to and manage crises.

Preparing for the unexpected is not the norm

In the past several decades, Mitroff and his colleagues have conducted numerous surveys of a great variety of organizations including American corporations, not-for-profits, governmental organizations, colleges and universities. We believe that K-12 educational institutions can learn from the general patterns and findings from these surveys.

Over decades, corporations, governmental organizations, and educational institutions have told us repeatedly that they are relatively more prepared for those crises or incidents that they have actually experienced and are less prepared for those crises or incidents they have yet to experience. The most frequently *experienced* crises are fires and explosions. The following types of crises were also *experienced* rather frequently: significant increases in costs; major crimes against faculty, staff, and students; major crimes committed by faculty, staff, and students; and major lawsuits. When they were asked to indicate how *prepared* they were for this same set of crises, the participants in the surveys generally felt they were most prepared for fires, major lawsuits, natural disasters, environmental disasters, major crimes against faculty, staff, and students and major crimes by faculty, staff, and students. The participants also indicated they were *least prepared* for: damage to their reputation; significant outbreaks of illness/disease among students, faculty, and staff; sabotage; terrorists' attacks; and food tampering. Not surprisingly, for the most part, these same crises were at the bottom of the list of those actually experienced by the institutions responding to the surveys.

In terms of the Jungian framework, these institutions have overemphasized a ST or reactive approach to Crisis Management. That is, they have prepared for the more frequent and the more familiar crises, but they have not prepared for unexpected crises.

Reactive Crisis Management or Risk Management is essentially a series of cost-benefit analyses that are grounded in historical data. These types of cost-benefit analyses tell us something about the expected costs of different crises. In probability theory, it is shown that the expected cost of a crisis (more generally an event of any sort) is equal to the probability of a crisis (event) happening multiplied by its expected cost.

DOI: 10.1057/9781137386045

Such analyses must be done very carefully because they suffer from several problems. First, reactive Crisis Management or Risk Management ignores the unexpected because it cannot be measured. It is hard to calculate the likelihood of those events that have not yet happened. As a result, crises that have never or very rarely happened before cannot be used in risk equations. Second, the cost of a crisis is as difficult to calculate as its probability. In fact, it is often impossible for affected stakeholders to agree even on the definition of a crisis, let alone its costs and consequences, and how much its consequences and preparation efforts will cost. In short, Risk Management or a ST approach to Crisis Management ignores the unexpected. In general, Risk Management tends to ignore events like 9/11 that although obviously high in costs and consequences are thought to be so low in probability that they can be "safely ignored."

The current financial/economic crisis is perhaps the most recent example of an unexpected crisis, or at least it was for most people. What is unexpected about the current financial crisis is not that it happened because financial crises have been with us for a long time, but its magnitude. A financial mega crisis such as the one that we experienced and of which we are still feeling the consequences is having different effects on schools than past recessions. It is forcing educational institutions to layoff instructors, implement furloughs, and increase student fees.

From the prospective of a school administrator, these findings are especially important. Although it is comforting to know that if an institution has experienced a particular crisis or incident they are thus better prepared for it next time, it is far less comforting to know that schools are not as well prepared for unexpected crises. Because educational institutions are much less prepared for unexpected or uncommon crises, such crises (reputation damage, significant outbreaks of illness/disease among students, faculty and staff, sabotage, terrorists' attacks, and food tampering) if and when they occur, are likely to cause significant problems for institutions.

There is another very important reason for preparing for those crises that a school has not yet experienced, and especially those that it thinks that it will never experience. *No crisis of which we are aware is ever a single, well-defined, and isolated crisis. Instead, all crises are a part of system of multiple crises.* That is, a system of *interdependent* crises typically occurs simultaneously, or one crisis sets off a chain reaction of other crises. In

DOI: 10.1057/9781137386045

other words, they are messy. Thus, one must not only be prepared for those crises that are most likely to occur, but also for those that have not yet occurred. So, should schools prepare for every and all crises that they can possibly experience? Is it possible to do this? Furthermore, should schools prepare for all crises they can envision and imagine? These questions are at the heart of Crisis Management.

What is crisis management?

Those involved in Crisis Management often make the mistake of looking at crises as well as preparation efforts for crises in isolation. It is critical for school leaders to recognize that Crisis Management is *not* thinking about or planning for a few or particular types of crises in isolation. Crisis Management *is* thinking about and planning for a wide range of crises and especially their interactions. Similarly, a Crisis Management program is not merely a set of emergency preparedness plans or procedures. Nor is it about preparing for a few isolated or particular types of crises. Crisis Management is a systematic way of planning for the interactions among various types of crises. It is often the events that occur because of the interactions between seemingly isolated incidents that turn a single, minor incident into a major crisis.

To make this process manageable and doable, Crisis Management first stresses that there are different types of crises (fires, natural disasters, loss of critical and/or confidential information, etc.) which cluster together in certain "families of crises." Next, it insists that *one plan for the occurrence of at least one crisis in each of the types or "families" of crises independently of their actual or hypothesized likelihood of occurrence.* In this way, one will have "covered the bases" as best as one can. Finally, one is then instructed to integrate crisis plans to take into consideration the distinct possibility of the simultaneous occurrence of multiple crises.

From the perspective of Crisis Management, it is essential that natural- and human-caused crises be viewed as integral and inseparable parts of the same continuous chain. In terms of the Jungian framework, this requires integrating T and F, that is, the technical and the social.

Even more critical is what natural disasters have taught us: the human-caused aspects or phases of a crisis (such as being prepared psychologically, having well-trained teams in place, etc.) *must precede* the natural phases! If the responses to the natural parts of the crisis are

DOI: 10.1057/9781137386045

not already in place before it occurs, then it will be too late to treat the human-caused aspects. For those in charge of Crisis Management, this means that as soon as a natural disaster occurs, then the responses to the human-caused aspects need to be immediate.

In fact, *current thinking is that all crises are human-caused.* There are certainly natural hazards such as earthquakes and hurricanes but natural hazards become crises only if in the case of an earthquake we build shoddy buildings, put profits above safety, fail to create a safety culture, etc. In other words, crises occur because of what humans and human designed and run institutions do or fail to do. In this sense, *there are NO natural disasters, only natural hazards which have the potential to become human-caused crises. In other words, humans, not Mother Nature, decide what to build and where.*

Hurricane Katrina was a disaster for the City of New Orleans, the State of Louisiana, and the entire country. The failure of the political and economic systems in New Orleans caused Katrina to become a disaster. For schools, it was the ripple effects of the hurricane, inadvertent and/or poor facilities management that led to the significant loss of financial, student, and research data. This could have been prevented or at the very least mitigated.

For instance, Tulane University's experience during and after Hurricane Katrina shows how difficult it is to plan fully for such interactions. As reported in the September 28, 2005 *Wall Street Journal,* Tulane was partially successful in dealing with one of the greatest crises due to natural hazards ever suffered in the United States. Nonetheless, there were still serious unanticipated consequences. According to the *Journal,* "When generators ran out of fuel, 27 massive freezers lost power, the resulting heat destroyed 33 years worth of blood samples collected as part of a research project into adolescent heart disease." One of the tenets of good Crisis Management is increasing the awareness of and planning for the overlap between facilities that are most at risk and those where significant research takes place.

No one disputes that it is not possible to prepare for every specific crisis or the interactions between them. However, our research and experience shows that the best-prepared organizations and schools have learned to form a crisis portfolio that is based on the crises identified below. As we indicated earlier, a robust crisis portfolio is formed by selecting *at least one* crisis from each of the various families for which to prepare. Here is a list of crises that can strike K-12 schools:

- Serious outbreaks of illness/disease
- Food tampering
- Employee, student sabotage
- Fires, explosions, and chemical spills
- Environmental disasters
- Significant drops in revenues
- Significant increase in cost
- Natural hazards
- Accidental loss of confidential or sensitive information and records
- Athletic scandals
- Theft or purposeful compromise of confidential or sensitive information and records
- Major lawsuits
- Terrorist attacks
- Damage to institutional reputation
- Ethical breaches by administrators/teachers/students
- Ethical breaches against administrators/teachers/students
- Other major crimes

Not only do the best prepared schools pay attention to those crises that they have already experienced, but also they plan for those that they have not yet experienced as well as the possible interactions between them. In other words, they take an NT or a proactive approach to Crisis Management. Specifically, they ask "what if" questions, and focus on future unknowns and the system as a whole. In sharp contrast to Risk Management, we cannot emphasize too much that proactive Crisis Management does not fixate on probabilities, measurement, or the agreement among affected stakeholders to define and prepare for crises. Rather, it accepts the fact that each affected stakeholder may have a different and conflicting yet legitimate view of crises. For proactive Crisis Management, "probabilities x consequences" is just one perspective or way to conceptualize crises. It is not the only way.

The main point of proactive Crisis Management is not to accomplish the impossible goals of preventing or being fully prepared for the unexpected. Rather, as much as is humanly possible, it is to learn and uncover, during the process of preparing for the unexpected, one's unanticipated vulnerabilities and blind spots. Table 9.1 summarizes the main differences between proactive and reactive Crisis Management. Notice that an institution can still be and should be proactive even in the response phase of a crisis.

DOI: 10.1057/9781137386045

TABLE 9.1 *Proactive and reactive crisis management behavior*[10]

	Preparation phase	Response phase
Reactive Crisis Management	Perform cost-benefit analyses, and prepare only for crises with high expected cost to the school Involve stakeholders in crisis preparations, only if mandated by law	If legally right, deny responsibility for the crisis and its effects on stakeholders; otherwise, admit some responsibility but fight it Comply when forced, and do only what is mandated by law
Proactive Crisis Management	Involve in crisis preparations a broader set of stakeholders than mandated by law Develop mutual trust and cooperation based relationships with all stakeholders Try to involve all stakeholders that may be harmed by the school's decisions and actions	Accept responsibility for the crisis Anticipate that the crisis may trigger a chain reaction of other crises Voluntarily attend to the needs of the victims, and tell the truth as you know it Get the worst about yourself out on your time before the media digit

Crisis typologies are useless unless...

In Chapter 3, we introduced the four main dimensions of the Jungian personality typology: (1) Introversion or I versus Extroversion or E; (2) Sensing S versus Intuition N; (3) Thinking T versus Feeling F; and (4) Perceiving P versus Judging J. We have then used the second and third dimensions to make sense of messes in general and The Education Mess in particular.

In this section, we want to use the first and the third dimensions to create an expanded typology of crises. We have already mentioned that introversion and extroversion refer to the source of a person's energy. In the context of crises, we refer to them as both the source of an organization' problems and the extent of its stakeholders' involvement. Specifically, we categorize crises in terms of whether they are caused by problems that are internal and/or external to an organization, and whether the consequences of such crises involve internal and/or external stakeholders. For example, a school building fire can be put in Quadrant 1 if it is caused by a technical problem such as an electrical short. It could be put in Quadrant 2 if the electrical short was caused by a natural hazard, say, an earthquake. If an arsonist caused the fire who was also a school employee or student employee then it could be put in Quadrant 4. If a terrorist who had targeted other schools caused the fire then the crisis could be put in Quadrant 3.

DOI: 10.1057/9781137386045

TABLE 9.2 *A crisis typology*

	Technical/Economic (T)		
	Quadrant 1 Loss of confidential information and records Fires, explosions Computer breakdown Bankruptcy	**Quadrant 2** Natural hazards Governmental crises State/city bankruptcies Financial/economic crises	
Internal (I)	**Quadrant 4** Athletic scandals Suicides School shootings by students or school employees False rumors, sick jokes by students or school employees Sexual harassment, abuse Employee, student sabotage	**Quadrant 3** Labor strikes Boycotts Terrorism False rumors, sick jokes by external agents School shootings by external agents Sabotage	*External (E)*
	Social/Personal (F)		

The crisis literature is full of typologies.[4] None of them, however, are complex enough to capture the full messiness of crises. All crises possess some properties that justify including them in any or all of the quadrants. As we have seen in the previous section, the human and the technical causes of crises are inseparable. Who is an internal or external stakeholder to the organization, and where to draw its boundaries is never clear. Are parents internal or external? How about the school's community, city, and state? Stakeholders can be and are simultaneously internal or external. Even if one doesn't live in Newtown, Connecticut, in the sense of NF, are we not all stakeholders?

So why create crisis typologies? Because the process of creating typologies and figuring out with what type(s) of crises we are dealing and who the stakeholders are helps us think and see connections that we would have otherwise missed. In other words, typologies are tools that help us reason. They are useless unless they are used as such.

Crisis Management Teams (CMTs)

Crisis Management Teams are a key component in preparing for and mitigating the effects of a crisis. While having a Crisis Management Team in place is of the upmost importance, by itself it is not enough.

DOI: 10.1057/9781137386045

The team must truly be a team, know how to work with one another, and the members must be aware of their individual as well as their collective responsibilities, and most important of all, their capabilities. This is best accomplished through regular meetings and training sessions. Ideally, all schools should have a Crisis Management Team that meets frequently and regularly.

The failure to adequately develop and train Crisis Management Teams inhibits an institution's ability to cope with crises. The establishment of a school-wide CMT is just the first step. To function effectively, the team must

- meet frequently on a regular basis;
- know how to work together;
- learn from the crises that occur in other schools and especially other very different type of institutions;
- assure that all significant incidents at school campus are fully analyzed and debriefed;
- be trained through the use of realistic simulations.

To be effective, CMTs teams must test Crisis Management plans and conduct crisis simulations annually at the very least. These simulations must focus on a broad range of crises and not just natural hazards. They must also be based on "worst case" scenarios, not unrealistic expectations. Simpson Scarborough, a team of crisis communications experts, found that the most frequent method for testing plans are "table top" simulations. Even more disturbing, they found that one-quarter of their respondents tested their plans less than once a year. In other words, organizations and institutions do not conduct crisis simulations as often as they need to do. Further, "when asked to identify their greatest fear should a crisis occur, 25% of the respondents said that they feared their crisis plan would not be executed properly, that people would be unaware of the plan, and that the plan has not been properly tested."[5]

How to collaborate

We cannot emphasize enough that crises are messy. Messy issues require collaboration. And collaboration is not easy. As we have noted throughout, people speak very different psychological languages, even when it

DOI: 10.1057/9781137386045

appears that they are all speaking the same native language. Below is a way of using the Jungian Personality Framework to create more effective collaboration between different types. This method is particularly useful in forming CMTs that are highly effective.

Crises are managed best before they happen; problems are solved best when they are small. But we acknowledge that it is hard to pick up the early warning signals to prevent an impending crisis, or to notice and solve small problems before they snowball into bigger ones.

One of the four Jungian types (ST, NT, NF, SF) is better than the others in picking up, that is, "sensing" and "feeling" signals: SFs. SFs excel at connecting with individuals, sensing and feeling their pain and suffering. Thus, they are excellent sources of information at the personal and individual levels. As much as they are good at sensing and feeling problems, SFs don't want to harm their deep, personal relationships. And, most of all, they dislike dealing with abstract concepts and ideas. Thus, at this point in the process, one needs help from NFs.

NFs not only speak the language of F, which is something they have in common with SFs, but they can also look at the more abstract, less grounded, bigger pictures to connect the dots. NFs' free-floating style helps them to excel at this task. To NFs, a bigger picture is a picture of larger communities. They are not only in tune with larger communities and their society, but they can motivate and mobilize greater numbers of people. NFs, however, prefer not to deal with the kind of impersonal logic required to synthesize their feelings and intuitions about the bigger problems at hand. Thus, one needs help from NTs.

NTs can not only speak the language of N, which is something they have in common with NFs, but they can also bring to the table the skills required to synthesize information from diverse sources. NTs' ability to look at things from multiple perspectives helps them excel at this task. To NTs, a bigger picture is a picture of interconnections within large systems. They can come up with multiple theories to explain and diagnose the problems; they can also propose multiple solutions to problems. NTs, however, prefer not to collect and analyze detailed data to test their intuitions and theories. Because they don't like to deal with details, they can make a lot of mistakes. Thus, at this point in the process, they need help from STs.

STs can not only speak the language of T, which is something they have in common with NTs, but they also prefer to deal with concrete data and facts. STs' impersonal style, and strong data collection and analysis

DOI: 10.1057/9781137386045

skills help them excel at this task. To STs, anything that is not grounded in observable facts are not real. Once problems are narrowed down to manageable sizes, STs test the feasibility of NTs' multiple models, and turn them into solutions concrete enough to implement in real life. STs, however, are not particularly fond of creating the kind of deep personal relationships that are necessary to turn their models into actionable solutions. It is exactly at this point in the process that one needs help again from SFs.

SFs can not only speak the language of S, which is something they have in common with STs, but they also prefer to connect with people at a the deepest level of being. It would seem that because SFs share S, it would make it easier and more effective for SFs to implement solutions suggested by STs. But unfortunately, it does not for their Ss are focused entirely differently. Whereas for SFs S is focused on individual people, S is focused on impersonal things for STs, thus making it hard for them to communicate. For this reason, SFs communicate better with NFs who can communicate with NTs. Thus, this rather complicated route is how SFs can eventually communicate with STs. Once again, the point is: All the types need one another.

Figure 9.1 shows that the collaboration process we have suggested is circular. As such, there is no clear beginning or an end to it. In the example above, we started with the SF quadrant, but we could have started with any of the other three quadrants.

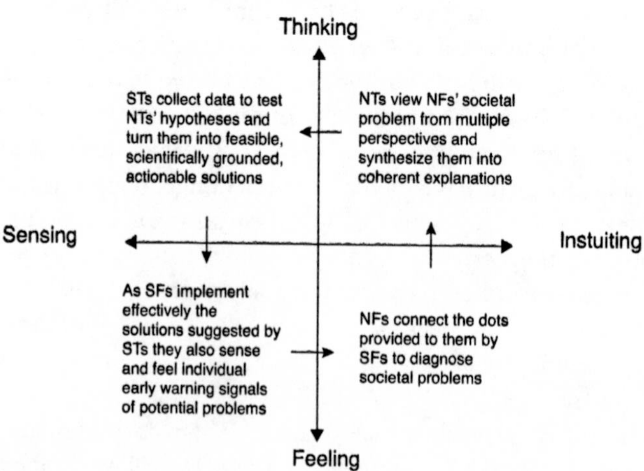

FIGURE 9.1 *How to collaborate*

DOI: 10.1057/9781137386045

The instant news environment

All organizations, whether business, not-for-profit or educational, now operate in a world of instant communications dominated by YouTube, Facebook, Twitter, the blogosphere, 24-hour news, instant analysis, and the demand for instant answers. For instance, the video of a University of Florida student who was shot by a Taser gun by campus police was one of the most viewed YouTube videos of 2007.

In this environment, it is critical that school leaders and administrators recognize that it is difficult, if not impossible, to contain or control completely the distribution of information about an incident or crisis. When an incident takes place on a school campus before traditional news journalists even arrive on campus, students may have already captured the event on their cell phones and sent pictures over the web. Nonetheless, traditional news journalists generally raise three types of questions:

1 What did you know about the incident/crisis and when did you know it?
2 If you didn't know about it, why not?
3 If you knew about it, why didn't you do something about it?

More often than not, questions about similar past events are raised and there is always an attempt to link past events to the current incident with the implication that the organization did not do what it could to have prevented the current situation. In other words, the media try to show that the current crisis is not an isolated aberration, but part of the systematic and systemic failure of the institution to plan adequately for crises and to address past deficiencies.

The shootings at the Virginia Tech are an unfortunate example of what not to do. Pictures of the shootings were all over the web before law enforcement or university officials could even respond to the second shooting. When news journalists arrived they quickly focused on two questions: 1) why it took so long to notify students about the first shooting and 2) once the identity of the shooter and his past medical history was made known, news journalists focused on why nothing was done to keep this student away from other students.

The Penn State child sex abuse scandal is another example of Crisis Management at its worst. According to the independent report by the law firm of Freeh Sporkin & Sullivan, LLP, the university's leaders, including

DOI: 10.1057/9781137386045

President Graham Spanier and the legendary coach Joe Paterno, did not do anything for more than a decade to protect the victimized children.[6] Instead, they tried to protect the reputations of the football team, its legendary coach, and university; for the failure to do so also meant less revenues, drop in enrollments and donor contributions as well as decreased state support.

A good Crisis Management plan must take into account the importance and consequences of the current instant news environment and be based on the following:

- It is virtually impossible to contain perfectly and thus put tight boundaries around any crisis;
- It is critical that a predetermined point person (or persons) be available to handle all communications;
- Because the blame game will begin almost immediately, a well-designed crisis communication plan must be in place so that it can be enacted instantly;
- All past incidents that are related even in the smallest way to the current crisis must be quickly and honestly disclosed.

Not paying attention to signal detection and ticking time bombs

A review of the decades long evidence on crises that schools and companies have experienced shows that most crises send out a trail of early warning signs that indicate that a crisis may be brewing or about to explode long before the crisis happens. If a school has the capability to pick-up these signals, transmit them to the right people, and take appropriate action, it is possible to prevent many crises before they occur or at least mitigate their adverse effects.

The effective use of signal detection, however, depends on two other important factors. Schools must understand the wide variety of activities they conduct everyday and identify the ticking time bombs associated with each of these activities.

Identifying Everyday Activities. In many ways, a school is as complex as business organizations. It not only provides education, but it also conducts a variety of activities. These activities may include facility and property management, health care and first aid, travel, educational

DOI: 10.1057/9781137386045

tours, food services, PTA meetings, special events, sports activities, competitions, budgeting, human resources, relations with the public, government and the foundation, community outreach programs, police and safety, and so on. Those in charge with overseeing the development and implementation of a Crisis Management plan for their school need to recognize that each of these activities involves different risks and if and when they interact with one another, the complexity of managing a crisis increases significantly.

Identifying Ticking Time Bombs. Ticking time bombs are potential crises that lie just under the surface of various educational and support activities that can easily be activated by exogenous or internal events. Table 9.3 lists various types of ticking time bombs that might be faced by a school. A CMT needs to review constantly these kinds of ticking time bombs to understand how they can be set off by various events and what actions need to be taken immediately to reduce the probability that they might turn into real crises for the school.

One of the most effective tools a CMT can use is what we call *crises maps.* These are maps of ticking time bombs that are meant to highlight important interactions among crises. To illustrate how this might work, a school can create a map that shows all the facilities and buildings on

TABLE 9.3 *Possible ticking time bombs at schools*

Criminal
- Rapes, Murders, Robberies
- Guns, Gangs, Terrorism

Informational
- Identity theft
- Violations of confidentiality
- Fraud

Building safety
- Substandard housing on and off campus
- Structural integrity
- Back-up generation
- Fires
- Appropriate evacuation routes

Athletics
- Recruiting practices
- Academic or sex scandals

Perceptual/Reputation
- False rumors, sick jokes, slander

Unethical behavior/Misconduct
- Fraud
- Plagiarism by students, faculty and administrators
- Record tampering
- Conflicts of interests by faculty, administrators and board **members**
- Negligence or failure to perform one's duty as a teacher or administrator

Financial
- Fraud
- Mismanagement, failure of a major unit
  Natural Hazards
- Weather related disasters, earthquakes

Health
- Disease outbreaks
- Epidemics
- Food safety and tampering

Legal/Labor Disputes
- Faculty strikes

DOI: 10.1057/9781137386045

its campus where confidential records and/or expensive equipment are located. Another map then highlights buildings that are most at risk for fire, natural hazards or could be compromised (e.g., lack of backup power or refrigeration). Next, these two maps are overlaid on each other. The buildings and facilities that overlap on both maps are the ones most at risk. If something like this had been done on the campuses in New Orleans, some of the research that was irretrievably lost in hurricane Katrina might have been saved. These maps can be used for all sorts of purposes such as to identify buildings where visitors and/or children most often visit, to highlight overlaps between computer facilities and building safety and to understand where student populations are clustered at different times during the day in case a school lockdown or evacuation is required.

Constant support for crisis management is critical

Most institutions "react" to a crisis in the sense that after a major crisis (such as the Sandy Hook High School Shooting), they allocate more resources to Crisis Management, but in time, with improved safety and long uneventful periods, the effects of the crisis are forgotten, institutions allocate less and less resources to crisis management. *These institutions confuse absence of failure with presence of safety.* As they lower margins of safety, they slowly "drift into failure".[7] Mitroff and Alpaslan surveyed a sample of the largest businesses in the United States about seven months before, three months after, one year after, and two years after 9/11. Most of these businesses, particularly if they felt they could be the target of a terrorist attack, increased their preparations significantly right after the attacks. One year later, they were still more prepared than they were before the attacks but less prepared than they were after the attacks. But two years later, they were only as prepared as they were before the attack.[8]

We cannot overemphasize enough the need for continuous, on-going, constant support for Crisis Management.

The belief system in place in many institutions that often makes it difficult to plan for a crisis

It is not only naïve but also dangerous for school leaders and administrators to believe that the school shootings or crises that strike other

DOI: 10.1057/9781137386045

schools are isolated events, and that such an event will not happen at their schools.

One of the most powerful detriments to effective Crisis Management is the belief system that exists at an institution. Does the institution believe that it is invulnerable to a wide variety of crises? Does it feel that it is worth the time and effort to be prepared for crises? Or does it have an attitude of *it can't happen here*?

Mitroff and his colleagues have found that (1) the more institutions believe they are invulnerable, the less they invest in Crisis Management and thus the more crisis-prone they become over time, and conversely, (2) the more vulnerable institutions believe they are, the more prepared they are, and the less vulnerable they become.[9]

Consider the attitudes and beliefs listed below:

"Crises can't and won't happen to us."
"Excellent organizations don't have crises."
"High profile organizations are more likely to experience major crises."
"It's impossible to prepare for every conceivable crisis."
"Crisis Management and crisis simulations are a waste of time and money."
"We are too busy to conduct crisis training."

If a majority of these beliefs are characteristic of an institution's culture, then it may be far more vulnerable and less prepared than it believes.

Recommendations

School leaders have significant responsibility for the on-going success of their institution. Ensuring that a school has a well developed, well-tested, broad-based Crisis Management plan is a critical aspect of fulfilling that responsibility. In this chapter, we have provided some current insights into how schools can prepare for crises and suggested a number of recommendations that are summarized below:

- Be prepared for a broad range of crises by developing a crisis typology and a crisis portfolio.
- Develop a list of activities in which the school is engaged.
- Develop a list of ticking time bombs and understand what exogenous and/or internal events could set them off.

DOI: 10.1057/9781137386045

- Develop a crisis map.
- Form a multidisciplinary crisis management team.
- Make sure that team is trained to handle a series of broad ranging and often unanticipated crises.
- Constantly train the team by simulating not only physical crises (earthquakes, fires) but also reputation crises.
- Make sure that the school has a clear chain of command for decision-making during a crisis, do this long before a crisis strikes.
- Make sure that adequate non-technology based communications are available.
- Find ways to increase support for crisis management as a leadership imperative.
- Systematically review and learn from the crises faced by other schools.

Leaders in schools have a significant fiduciary responsibility to their institutions and the stakeholders of those institutions. In today's world, ensuring the development, implementation, and effective functioning of a crisis management team and plan is an essential part of that fiduciary duty. Implementing the recommendations outlined above will help make Crisis Management a priority of top management and help create an environment that mitigates crises that are bound to occur, if not inevitable in today's messy world.

Notes

1 Pearson, C. M. and Clair, J. A. 1998. "Reframing Crisis Management," *Academy of Management Review*, Vol. 23, No. 1, pp. 59–77.

2 Mitroff, I. I., Alpaslan, C. M., and Green, S. E. 2004. "Crises as Ill-Structured Messes: Philosophical Issues of Crisis Management," *The International Studies Review*, Vol. 6, No. 1, pp. 175–182.

3 Coleman, L. and Helsloot, I. (2007). "On the need for quantifying corporate crises and other man-made disasters," *Journal of Contingencies and Crisis Management*, Vol. 15, No. 3, pp. 119–122; Mitroff, I. I. and Alpaslan C. M. (2003). "Preparing for Evil," *Harvard Business Review*, Vol. 81, No. 4, pp. 109–115; Pearson, C. M. and Mitroff, I. I. (1993) "From Crisis-Prone to Crisis-Prepared," *Academy of Management Executive*, Vol. 7, No. 1, pp. 48–59; Runyan, R.C. (2006) "Small Business in the Face of Crisis: Identifying Barriers to Recovery from a Natural Disaster," *Journal of Contingencies and Crisis*

DOI: 10.1057/9781137386045

Management, Vol. 14, No. 1, pp. 12–26; Sheaffer, Z., and Mano-Negrin, R. (2003) "Executives' Orientations as Indicators of Crisis Management Policies and Practices," *Journal of Management Studies*, Vol. 40, No. 2, pp. 573–606.

4 Marcus, A. A. and Goodman, R. S. 1991, "Victims and Shareholders: the Dilemma of Presenting Corporate Policy during a Crisis," *Academy of Management Journal*, Vol. 34, No. 2, pp. 281–305; Mitroff, I. I. and Alpaslan, C. M. 2003, "Preparing for Evil," *Harvard Business Review*, Vol. 81, No. 4, pp. 109–115; Pearson, C. M. and Mitroff, I. I. 1993, "From Crisis-Prone to Crisis-Prepared," *Academy of Management Executive*, Vol. 7, No. 1, pp. 48–59; Gundel, S. (2005). "Towards a New Typology of Crises," *Journal of Contingencies and Crisis Management*, Vol. 13, No.3, pp. 106–115.

5 Simpson Scarborough, White Paper, "When Disaster Strikes: How Well Are Colleges Prepared for a Crisis" October 4, 2007, www.simpsonscarborough.com/documents/CollegeCrisisPreparednessWhitePaper.pdf

6 "Penn State's Part," *The New York Times*. July 12, 2012. Retrieved Feb 1, 2013.

7 Dekker, Sidney (2011), *Drift into Failure: From Hunting Broken Components to Understanding Complex Systems*, Ashgate, Aldershot.

8 Mitroff and Alpaslan, "Preparing for Evil"

9 Mitroff, I. I., Shrivastava, P., and Udwadia Firdaus E. (1987). "Effective Crisis Management," *Academy of Management Executive*, Vol. 1, No. 3, pp. 283–292.

10 Alpaslan, C. M. (2009) "Ethical Management of Crises: Shareholder Value Maximization Or Stakeholder Loss Minimization?" *Journal of Corporate Citizenship*, Vol. 36(Winter), pp. 41–50; Alpaslan, C. M., S. Green, and I. Mitroff (2009) "Corporate Governance in the Context of Crises: towards a Stakeholder Theory of Crisis Management, *Journal of Contingencies and Crisis Management*, Vol. 17, No. 1, pp. 38–49; Shrivastava, P. 1993. "Crisis Theory/Practice: towards a Sustainable Future," *Industrial and Environmental Crisis Quarterly*, Vol. 7, pp. 23–42; Jawahar, I. M. and McLaughlin, G. L. 2001. "Toward a Descriptive Stakeholder Theory: An Organizational Life Cycle Approach," *Academy of Management Review*, Vol. 26, No. 3, pp. 397–414; Clarkson, M. B. E. 1995. "A stakeholder framework for analyzing and evaluating corporate social performance," *Academy of Management Review*, Vol. 20, pp. 92–117; McAdam, T. W. 1973. "How to Put Corporate Social Responsibility into Practice,". *Business and Society Review/Innovation*, Vol. 6, pp. 8–16.

DOI: 10.1057/9781137386045

Epilogue

Mitroff, Ian I., Hill, Lindan B., and Alpaslan, Can M. *Rethinking the Education Mess: A Systems Approach to Education Reform*. New York: Palgrave Macmillan, 2013.
DOI: 10.1057/9781137386045.

DOI: 10.1057/9781137386045

What is a system?

Recall that a system is an *intentionally designed, systematically organized, whole entity* that has one or more *essential functions* so that an individual and/or group of people are thereby able to realize their *purposes.*

A system consists of *at least two or more* essential parts without which it cannot accomplish its defining functions. And a system's essential parts do not have an independent effect on the whole system.

Improvement in the parts taken separately does not necessarily improve a system overall as a whole. Indeed, it often leads to its failure and complete destruction.

A system has defining (emergent) properties that none of its parts has.

What is a Mess?

A *mess* is a *whole system of problems* that is poorly organized, even disorganized.

A mess consists of at least two different problems. *None* of the problems that constitute a mess even *exists* (and thus can be understood) independently of all of the other problems that are integral parts of the mess. *At least one* of these problems is a wicked problem, a problem that cannot be completely defined, let alone "solved," by any known academic discipline or profession either solely by itself or in combination with the others.

A mess has no subset of problems that has an independent effect on the whole mess.

A mess as a whole has defining (emergent) properties that none of the "individual elements" have.

In short, a mess is a system at a higher level of complexity.

It is a "messier system" for the following reasons.

A mess cannot be defined independently of all the stakeholders who affect and are affected by the mess. There are no neutral terms in a mess. Ideology plays a central role. All terms are emotionally or morally loaded to some degree,

A mess contains the anxieties, dreams, emotions, fears, hopes, and accompanying assumptions, beliefs, and myths, both conscious and unconscious, of its stakeholders. A mess also contains the memories of previous attempts, successful and otherwise, to manage the mess. In

DOI: 10.1057/9781137386045

short, a mess *potentially* contains everything pertaining to the human condition.

In principle, messes are related to and a part of every other mess such that every mess contains at least one element (one problem, one powerful underlying emotion, one deep assumption/belief, etc.) from at least one other mess.

Messes are like fractals. They are "messy" all the way down and all the way up. They do not begin or end at any particular level of "reality."

Not all messes are crises but messes always contain a high potential for crises.

One cannot "solve" messes. One must manage (cope with) messes as best one can.

Some of key heuristics for avoiding Type 3 errors

First of all, never ever trust a single formulation of a mess.

Always view a mess from as many different perspectives as possible such as psychological, sociological, anthropological, historical, moral, political, technological, financial, and spiritual perspective, among many others.

Get as many different stakeholders as possible to formulate a mess.

Seek out and sweep in the analyses of those experts who cross-connect fields.

Monitor, examine, and challenge taken-for-granted assumptions and beliefs. In particular, examine the deep and thereby often unconscious assumptions that are made about different stakeholders and the mess in hand.

Investigate and understand the complexity of interactions. Design scenarios that deliberately probe for difficult interactions. Give special attention to the most improbable interactions, the least important interactions, the most damaging interactions, the most counter-intuitive, paradoxical, and unintended interactions within and between messes.

Examine extremes. Pay special attention to outliers.

Never accept conventional, traditional constraints or boundaries. Instead perturb the ordinary/conventional. Rock the boat. Ask "smart-dumb" questions. Play the devil's advocate.

In sum, throughout this book, we have emphasized that a "mess" is a whole system of problems that are so strongly interconnected such that

DOI: 10.1057/9781137386045

they can't even be stated (formulated) properly, let alone dissolved and/or resolved, independently of one another. Strongest of all, none of the problems even exists independently of all the other problems that constitute a mess. To take any problem out of a mess is not only to distort the nature of the problem, but the entire mess. To "manage a mess" is to "manage its 'critical interactions.'" But, what's "critical" can be ascertained only by looking at the entire mess.

Most important of all, we also presented a series of heuristics for coping with and managing messes. If there weren't any means whatsoever for coping with messes, then there would no case for hope.

A supreme spiritual challenge

In the end, the ability to manage messes is ultimately a spiritual challenge. It is certainly not a technical challenge alone.

To our knowledge, no one has posed the nature of this challenge better than William James, one of the founders of the distinct brand of American philosophy known as Pragmatism. In addition, James is arguably one of America's greatest philosophers, if not its greatest.

Since the beginnings of systems thinking and messes can be traced back to James, it is more than fitting that we end with him.

In some of the most powerful and inspirational words ever written, James put a supreme challenge to humankind. We quote:

> ... Suppose that the world's author put the [following] case to you before [the exact moment of] creation, saying: "I am going to make a world not certain to be saved, a world the perfection of which shall be conditional merely, the condition being that each several agent [stakeholder] does its own 'level best.' I offer you the chance of taking part in such a world. Its safety, you see, is unwarranted. It is a real adventure with real danger, yet it may win through. It is a social scheme of co-operative work genuinely to be done. Will you join the procession? Will you trust yourself and trust the other agents enough to face the risk?"[1]

With little change in wording, James' inspirational thoughts apply equally to the situation in which we find ourselves:

> On the one hand, I [The Supreme Creator] offer you a world that is simple and completely determined. It is a world that is governed by complete and exact causes and effects. In principle, everything is knowable. In this world, your actions will make little if any difference, especially since everything

DOI: 10.1057/9781137386045

is predetermined. On the other hand, I offer you a world of enormous complexity, in short of world of complex, messy systems. In this world, which is not guaranteed to work perfectly, your actions will not only make a difference, but are key if one is to make any differences whatsoever. That is, your actions are essential if in principle there is to be the possibility of making any differences. But to do so, you will have to learn to cope with and manage messes as best you can. And, you have to learn to do something even more challenging. You will not only have to develop the energy and ability to tolerate messes, but you will have to develop the will to appreciate them with all your heart, soul, and being. I urge you to choose carefully and wisely.

Note

1 James, William, *Pragmatism*, Prometheus Books, Buffalo, New York, 1991, p. 129.

DOI: 10.1057/9781137386045

Bibliography

Ackoff, Russell L., *Re-Creating the Corporation*, Oxford University Press, USA, 1999.

Ackoff, Russell L. and Greenberg, Daniel, *Turning Learning Right Side Up, Putting Education Back On Track*, University of Pennsylvania Press, Philadelphia, 2003.

Ackoff, Russell L. and Rodin, Sheldon, *Redesigning Society*, Stanford University Press, 2003.

Alon, Nahi, and Omer, Haim, *The Psychology of Demonization: Promoting Acceptance and Reducing Conflict*, Routledge, New York, 2006.

Alpaslan, Can M. "Ethical Management of Crises: Shareholder Value Maximization or Stakeholder Loss Minimization?" *Journal of Corporate Citizenship*, Vol. 36(Winter), 2009.

Alpaslan, Can M., Green, Sandy, and Mitroff, Ian. "Corporate Governance in the Context of Crises: towards a Stakeholder Theory of Crisis Management", *Journal of Contingencies and Crisis Management*, Vol. 17, No. 1, 2009, pp. 38–49.

Blow, Charles,, "Revolutionary Language," *The New York Times*, January 12, 2013, p. A17.

Christen, Clayton, Horn, Michael, and Johnson, Curtis, *Disrupting Class, How Disruptive Innovation Will Change the Way the World Learns*, McGraw Hill, New York, 2008.

Churchman, C. West, *The Design of Inquiring Systems: Basic Concepts of Systems and Organizations*, Basic Books, New York, 1971.

DOI: 10.1057/9781137386045

Clarkson, Max B. E., "A Stakeholder Framework for Analyzing and Evaluating Corporate Social Performance," *Academy of Management Review*, Vol. 20, 1995, pp. 92–117.

Coleman, L. and Helsloot, I. 2007. "On the Need for Quantifying Corporate Crises and Other Man-made Disasters," *Journal of Contingencies and Crisis Management*, Vol. 15, No. 3, pp. 119–122.

Cuban, Larry, and Usdan, Michael, *Powerful Reforms with Shallow Roots: Improving America's Urban Schools*, Teachers College Press, New York, 2003.

Datnow, Amanda, *Integrating Educational Systems for Successful Reform in Diverse Contexts*, Cambridge University Press, New York, 2006.

Datnow, Amanda, Lasky, Sue, Stringfield, Sam and Teddlie, Charles, *Integrating Educational Systems for Successful Reform in Diverse Contexts*, Cambridge University Press, New York, 2006.

Dekker, Sidney, *Drift into Failure: From Hunting Broken Components to Understanding Complex Systems*, Ashgate, Aldershot, 2011.

Demausse, Llyod, *Foundations of Psychohistory*, Creative Roots, Inc., New York, 1982.

Dobie, Will, "Are High Quality Schools Enough to Close the Achievement Gap? Evidence from a Social Experiment In Harlem," National Bureau of Economic Research, Working Paper 15473, Cambridge, MA, November 2009.

Dobbie, Will and Fryer, Roland, "Are High Quality Schools Enough to Close the Achievement Gap? Evidence from a Social Experiment in Harlem," *National Bureau of Economic Research*, Cambridge, MA, 2009.

Dolnick, Edward, *The Clockwork Universe: Isaac Newton, the Royal Society, and the Birth of the Modern World*, Harper, New York, 2011.

Duke, Daniel L., "Tinkering and Turnarounds: Understanding the Contemporary Campaign to Improve Low-Performing Schools," in Stuit, David, and Stringfield, Sam (eds), "Special Issue: Responding to the Chronic Crisis in Education: The Evolution of the School Turnaround Mandate," *Journal of Education of Students Placed at Risk*, Vol. 17, Nos 1–2, January–June, 2012, pp. 9–24.

Duncan, Greg J. and Murnane, Richard J. (eds), *Whither Opportunity: Rising Inequality, Schools, and Children's Life Chances*, Russell Sage Foundation, New York, 2011.

Foroohar, Rana, "These Schools Mean Business," *Time*, April 9, 2012, p. 26.

DOI: 10.1057/9781137386045

Gharajedaghi Jamshid, *A Prologue to National Development Planning*, Greenwood Press, New York, 1986.

Gharajedaghi Jamshid, *Systems Thinking, Managing Chaos and Complexity*, BH, Elsevier, Boston, 2006.

Gilligan, Carol, *In A Different Voice*, Harvard University Press, Cambridge, MA, 1982.

Gilligan, James, *Why Some Politicians Are More Dangerous than Others*, Polity, Cambridge, UK, 2011.

Gladwell, Malcolm, *Outliers: The Story of Success*, Little, Brown and Company, Boston, 2008.

Grossman, Klaus, Grossmann, Karin, and Waters, Everett (eds), *Attachment from Infancy to Adulthood: The Major Longitudinal Studies*, Guilford, New York, 2005.

Grubbs, F. E., "Procedures for Detecting Outlying Observations in Samples," *Technometrics*, Vol. 11, 1969, pp. 1–21.

Gundel, S., "Towards a New Typology of Crises," *Journal of Contingencies and Crisis Management*, Vol. 13, No.3, 2005, pp. 106–115.

Haidt, Jonathan, *The Righteous Mind, Why Good People Are Divided by Politics and Religion*, Pantheon, New York, 2012.

Hansen, Michael, "Key Issues in Empirically Identifying Chronically Low-Performing and Turnaround Schools," in Stuit, David and Stringfield, Sam (eds), "Special Issue: Responding to the Chronic Crisis in Education: The Evolution of the School Turnaround Mandate," *Journal of Education of Students Placed at Risk,* Vol. 17, Nos 1–2, January–June, 2012, pp. 55–56.

Heck, W. P., The School Shooter: One Community's Response. *FBI Law Enforcement Bulletin*, Vol. 70, No. 9, 2001, pp. 9–13.

Henigan, Dennis, *Lethal Logic: Exploding the Myths that Paralyze American Gun Policy*, Potomac Books, Washington, DC, 2009, p. 35.

James, William, *Pragmatism*, Prometheus Books, Buffalo, New York, 1991, p. 129.

Jawahar, I. M. and McLaughlin, G. L., "Toward a Descriptive Stakeholder Theory: An Organizational Life Cycle Approach," *Academy of Management Review*, Vol. 26, No. 3, 2001, pp. 397–414.

Jennings, Jack, "Reflections on a Half-Century of School Reform: Why We Have Fallen Short and Where Do We Go From Here?", Center on Education Policy, Washington, DC, January, 2012.

Johnson, Susan Moore, *Finders and Keepers, Helping New Teachers Survive and Thrive in Our Schools*, Jossey-Bass, San Francisco, 2004.

DOI: 10.1057/9781137386045

Jung, Carl, *Psychological Types*, Vol. Six, *Collected Works*, Princeton University Press, 1971.

Karen, Robert, *Becoming Attached: Unfolding the Mystery of the Infant-Mother Bond and Its Impact on Later Life*, Warner Books, New York, 1994.

Klein, Joel, and Rice, Condoleeza, Chairs, "U.S. Educational Reform and National Security," *Independent Task Force Report No. 68*, Council on Foreign Relations, New York, p. viii.

Klein, Melanie, *The Psychoanalysis of Children*, Nabu Public Domain Reprints, Grove Press, New York, 1960.

Lakoff, George, *Moral Politics*, University of Chicago, 1996.

Langman, Peter, *Why Kids Kill*, Palgrave Macmillan, 2010.

Lebrun, Marcel, Books, *Blackboards, and Bullets: School Shootings and Violence in America*, Rowland & Littlefield Education, Lanham, MD, 2009.

Lessig, Lawrence, *Republic Lost*, Twelve, New York, 2011.

Levin, Kelly, Cashore, Benjamin, Bernstein, Steven and Auld, Graeme. "Overcoming the Tragedy of Super Wicked Problems: Constraining Our Future Selves to Ameliorate Global Climate Change," *Policy Science*, Vol. 45, 2012, pp. 123–152.

Loveless, Tom (ed.), *Conflicting Missions, Teachers Unions and Educational Reform*, Brookings, Washington, DC, 2000.

Marcus, A. A. and Goodman, R. S., "Victims and Shareholders: The Dilemma of Presenting Corporate Policy during a Crisis," *Academy of Management Journal*, Vol. 34, No. 2, 1991, pp. 281–305.

Mathew, David, *Is There a Public for Public Schools?*, Kettering Foundation Press, Dayton, OH, 1996, p. 48.

McAdam, T. W., "How to Put Corporate Social Responsibility into Practice,". *Business and Society Review Innovation*, Vol. 6, 1973, pp. 8–16.

McIntire, Mike, "Selling a New Generation on Guns," *The New York Times*, Sunday, January 27, 2013, p. 1.

McKenna, Maryn "Clean Sweep: Hospitals Bring Janitors to the Front Lines of Infection Control," *Scientific American*, September 2012, pp. 30–31.

Merseth, Katherine, *Inside Urban Charter Schools: Promising Strategies in Five High-Promising Schools*, Harvard Education Press, Cambridge, MA, 2010.

Metaxas, Eric, *Amazing Grace: William Wilberforce and the Heroic Campaign to End Slavery*, Harper Row, New York, 2007.

DOI: 10.1057/9781137386045

Mitroff, Ian I. "The Complete and Utter Failure of Traditional Thinking in Comprehending the Nuclear Predicament, Why It's Impossible to Formulate A Paradox Free Theory of Nuclear Policy," *Technological Forecasting and Social Change*, Vol. 29, 1986, pp. 51–72.

Mitroff, Ian I. and Alpaslan, Can M., "Preparing for Evil," *Harvard Business Review*, Vol. 81, No. 4, 2003, pp. 109–115.

Mitroff, Ian I., and Silvers, Abe, *Dirty Rotten Strategies: How We Trick Ourselves and Others into Solving the Wrong Problems Precisely*, Stanford University Press, Palo Alto, CA, 2009.

Mitroff, Ian I., Alpaslan, Can M., and Green, Sandy E., "Crises as Ill-Structured Messes: Philosophical Issues of Crisis Management," *The International Studies Review*, Vol. 6, No. 1, 2004, pp. 175–182.

Mitroff, Ian. I., Shrivastava, P., and Udwadia Firdaus E., "Effective Crisis Management," *Academy of Management Executive*, Vol. 1, No. 3, 1987, pp. 283–292.

Neguera, Pedro, "Stretching the School Safety Net," *The Nation*, January 2, 2012, p. 23.

Pearson, Chris M., and Clair, J. A., "Reframing Crisis Management," *Academy of Management Review*, Vol. 23, No. 1, 1998, pp. 59–77.

Pearson, Chris M., and Mitroff, I., "From Crisis-Prone to Crisis-Prepared," *Academy of Management Executive*, Vol. 7, No. 1, 1993, pp. 48–59.

Postman, Neil, *The End of Education: Redefining the Value of School*, Vintage Books, New York, 1996, p. 172.

Ravitch, Diane, *The Life and Death of the Great American School System: How Testing and Choice Are Undermining Education*, Basic Books, New York, 2010

Robin, Corey, *The Reactionary Mind*, Oxford University Press, New York, 2011, p. 215.

Rothstein, Richard, *Grading Education: Getting Accountability Right*, EPI, 2008.

Runyan, R.C., "Small Business in the Face of Crisis: Identifying Barriers to Recovery from a Natural Disaster," *Journal of Contingencies and Crisis Management*, Vol. 14, No. 1, 2006, pp. 12–26.

Sheaffer, Z., and Mano-Negrin, R., "Executives' Orientations as Indicators of Crisis Management Policies and Practices," *Journal of Management Studies*, Vol. 40, No. 2, 2003, pp. 573–606.

Shrivastava, Paul, "Crisis Theory/Practice: towards a Sustainable Future," *Industrial and Environmental Crisis Quarterly*, Vol. 7, 1993, pp. 23–42.

DOI: 10.1057/9781137386045

Smith, Christian, *Moral Believing Animals*, Oxford University Press, New York, 2003.

Smith, E.O., *When Culture and Biology Collide*, Rutgers University Press, New Brunswick, NJ, 2002.

Sorensen, Geog, *A Liberal World Order in Crisis*, Cornell University Press, 2011.

Spillhaus, Elizabeth, and O'Shaughnessy, Edna (eds), *Projective Identification: The Fate of a Concept*, Routledge, London, 2012.

Stuit, David, and Stringfield, Sam (eds), "Special Issue: Responding to the Chronic Crisis in Education: The Evolution of the School Turnaround Mandate," *Journal of Education of Students Placed at Risk*, Vol. 17, Nos 1–2, January–June, 2012.

Thernstrom, Abigal and Thernstrom, Stephan, *No Excuses: Closing the Racial Gap in Learning*, Simon & Schuster, New York, 2003.

Tough, Paul, *Whatever It Takes, Geoffrey Canada's Quest to Change Harlem and America*, Mariner, New York, 2009.

Toulmin, Stephen, *The Uses of Argument*, Cambridge University Press, Cambridge, UK, 1958.

Tully, Matthew, *Searching for Hope: Life at a Failing School in the Heart of America*, Indiana University Press, Bloomington, 2012, p. 18.

Waddock, Sandra, *Not By Schools Alone, Sharing Responsibility for America's Education Reform*, Praeger, Westport, CN, 1995.

Warnick, Bryan, Johnson, Benjamin, and Rocha, Samuel, "Tragedy and the Meaning of School Shootings," *Educational Theory*, Vol. 60, No. 3, 2010, pp. 371–390.

DOI: 10.1057/9781137386045

Index

DOI: 10.1057/9781137386045

DOI: 10.1057/9781137386045

DOI: 10.1057/9781137386045

DOI: 10.1057/9781137386045

DOI: 10.1057/9781137386045

CPSIA information can be obtained at www.ICGtesting.com
Printed in the USA
LVOW08*1836151113

361401LV00006B/157/P